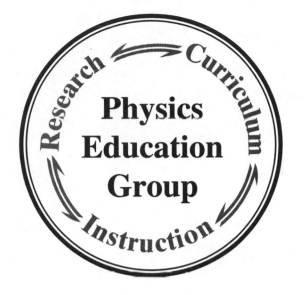

Physics Education Group

Research — Curriculum — Instruction

Tutorials in Introductory Physics

Preliminary Edition

Lillian C. McDermott, Peter S. Shaffer
and the Physics Education Group

Department of Physics
University of Washington

Prentice Hall
Upper Saddle River, New Jersey 07458

Prentice Hall Series in Educational Innovation

ACQUISITION EDITOR: *Alison Reeves*
PRODUCTION EDITOR: *Alison Aquino*
MANUFACTURING MANAGER: *Trudy Pisciotti*
COVER DESIGNER: *DeFranco Design, Inc.*

 © 1998 by Prentice-Hall, Inc.
Simon & Schuster/A Viacom Company
Upper Saddle River, New Jersey 07458

Printed in the United States of America
10 9 8 7 6 5 4 3 2

ISBN   0-13-954637-5

Prentice-Hall International (UK) Limited, *London*
Prentice-Hall of Australia Pty. Limited, *Sydney*
Prentice-Hall Canada Inc. *Toronto*
Prentice-Hall Hispanoamericana, S.A., *Mexico*
Prentice-Hall of India Private Limited, *New Delhi*
Prentice-Hall of Japan, Inc., *Tokyo*
Simon & Schuster Asia Pte. Ltd., *Singapore*
Editora Prentice-Hall do Brasil, Ltda., *Rio de Janeiro*

Preface

Tutorials in Introductory Physics is a set of instructional materials intended to supplement the lectures and textbook of a standard introductory physics course. The emphasis in the tutorials is not on solving the standard quantitative problems found in traditional textbooks, but on the development of important physical concepts and scientific reasoning skills.

There is increasing evidence that after instruction in a typical course, many students are unable to apply the physics formalism that they have studied to situations that they have not expressly memorized. In order for meaningful learning to occur, students need more assistance than they can obtain through listening to lectures, reading the textbook, and solving standard quantitative problems. It can be difficult for students who are studying physics for the first time to know what they do and do not understand and to learn to ask themselves the type of questions necessary to come to a functional understanding of the material. *Tutorials in Introductory Physics* provides a structure that promotes the active mental engagement of students in the process of learning physics. Questions in the tutorials guide students through the reasoning necessary to construct concepts and to apply them in real-world situations. The tutorials also provide practice in interpreting various representations (*e.g.,* formulas, graphs, diagrams, verbal descriptions) and in translating back and forth between them. For the most part, the tutorials are intended to be used after concepts have been introduced in the lectures and the laboratory, although most can serve to introduce the topic as well.

The tutorials comprise an integrated system of pretests, worksheets, and homework assignments. The tutorial sequence begins with a pretest. These are usually on material already presented in lecture but not yet covered in tutorial. The pretests help students identify what they do and not understand about the material and what they are expected to learn in the upcoming tutorial. They also inform the instructors about the level of student understanding. During a tutorial session, students work together on worksheets that provide the structure for these sessions. The worksheets consist of carefully sequenced tasks and questions. Students are expected to construct answers for themselves through discussions with their classmates and with tutorial instructors. The tutorial instructors do not lecture but ask questions designed to help students find their own answers. The tutorial homework reinforces and extends what is covered in the worksheets. For the tutorials to be most effective, it is important that course examinations include qualitative questions that emphasize the concepts and reasoning skills developed in the tutorials.

This preliminary edition of *Tutorials in Introductory Physics* contains a sufficient number of tutorials for a one-year course. Subsequent editions will include material on additional topics. The goal is to have enough tutorials to allow for two tutorial sessions each week.

The tutorials are designed for a small class setting that enables students to work together in groups of 3-4. However, in many cases, the materials can be adapted for use as interactive tutorial lectures. The tutorials are equally appropriate for calculus- and algebra-based courses.

Tutorials in Introductory Physics has been developed and tested at the University of Washington and pilot-tested at other colleges and universities.

Acknowledgments

Tutorials in Introductory Physics is the product of close collaboration by many members of the Physics Education Group at the University of Washington. Since 1991, when development of this curriculum began, significant contributions have been made by: Bradley Ambrose, Chris Border, Andrew Boudreaux, Patricia Chastain, Gregory Francis, Randal Harrington, Paula Heron, Stephen Kanim, Christian Kautz, Pamela Kraus, Michael Loverude, Graham Oberem, Luanna Gomez Ortiz, Tara O'Brien Pride, Christopher Richardson, Rachel Scherr, Mark Somers, Richard Steinberg, Stamatis Vokos, and Karen Wosilait, with special recognition for her work on the tutorials on waves and optics. The editorial assistance of Joan Valles in the preparation of this preliminary edition is deeply appreciated.

The collaboration of our colleagues in the Physics Department has been invaluable. Faculty who have been lecturers in the introductory calculus-based sequence, and graduate and undergraduate students who have served as tutorial instructors have made many useful comments. Among the students who have worked with us, Chris Becker and Scott Randol have been especially helpful.

Contributions have also been made by long-term visitors to our group. Since 1991, these have included: John Howell (Earlham College), Brian McInnes (University of Sydney), Daryl Pedigo (Austin Community College), E.F. (Joe) Redish (University of Maryland), Edwin Taylor (Boston University), and Walter Wales (University of Pennsylvania). Joe's enthusiasm for the project provided a major incentive for the publication of this Preliminary Edition. Physics instructors who have pilot-tested the tutorials and have provided valuable feedback include: Edward Adelson (The Ohio State University), John Christopher (University of Kentucky), James Freericks and Amy Liu (Georgetown University), Gregory Kilcup (The Ohio State University), Eunsook Kim (Seoul National University), Eric Mazur (Harvard University), James Poth (Miami University), E.F. Redish (University of Maryland), Jeff Saul (Prince George's Community College), and Beth Thacker (Grand Valley State University).

We thank our editor, Alison Reeves, for her encouragement and advice. We also gratefully acknowledge the support of the National Science Foundation, which has enabled the Physics Education Group to conduct the ongoing, comprehensive program of research, curriculum development, and instruction that has produced *Tutorials in Introductory Physics*. The tutorials have benefited from the past and concurrent development of *Physics by Inquiry* (©1996 John Wiley & Sons, Inc.), our other NSF-funded curriculum development project. *Tutorials in Introductory Physics* and *Physics by Inquiry* share a common research base and portions of each have been adapted for the other. We appreciate the cooperation of John Wiley & Sons, Inc. in facilitating this preliminary edition of *Tutorials in Introductory Physics*.

Table of Contents

Part I: Mechanics

Kinematics

Newton's laws

Energy and momentum

Rotation

Part II: Electricity and magnetism

Electrostatics

Electric circuits

Magnetism

Electromagnetism

Part III: Waves

Part IV: Optics

Geometrical optics

Physical optics

Kinematics

Tutorials in Introductory Physics
McDermott, Shaffer, & P.E.G., U.Wash.

©Prentice Hall
Preliminary Edition, 1998

I.   Motion with constant velocity

Each person in your group should obtain a ruler and at least one ticker tape segment from the staff. All the tape segments were generated using the same ticker timer. Do not write on or fold the tapes. If a ticker timer is available, examine it so that you are familiar with how it works.

A. Describe the motion represented by your segment of tape. Explain your reasoning.

B. Compare your tape with those of your partners.

How does the time taken to generate one of the short tape segments compare to the time taken to generate one of the long tape segments? Explain your answer.

Describe how you could use your answer above to arrange the tapes in order by speed.

C. Suppose the ticker timer that made the dots strikes the tape every $1/60^{th}$ of a second.

How far did the object that generated your tape segment move in: $1/60^{th}$ of a second? $2/60^{th}$ of a second? $3/60^{th}$ of a second? Explain your answer.

Predict how far the object would move in: 1 second, $1/120^{th}$ of a second. Explain the assumption(s) you used to make your predictions.

D. In your own words, describe the procedure you would use to calculate the speed of an object.

Determine the speed of the object that generated each of your tapes. Record your answers below.

Give an interpretation of the speed of the object, *i.e.*, explain the meaning of the number you just calculated. Do not use the word "speed" in your answer. (*Hint:* Which of the distances that you calculated in part C is numerically equal to the speed?)

Write the speed of the object that generated each tape on a small piece of paper and attach the paper to the tape. Express your answer in terms of centimeters and seconds.

E.   A motion that generates a sequence of evenly-spaced dots on a ticker tape is called motion with *constant velocity*. Explain the assumption about the motion that is being made when this phrase is applied.

   Discuss with your partners whether the object that generated your tape was moving with constant velocity.

F.   A model train moving with constant velocity travels 60 cm for every 1.5 s that elapses. Answer the questions below and discuss your reasoning with your partners.

   1.   Is there a name that is commonly given to the quantity represented by the number 40? (40 = 60/1.5) If so, what is the name?

      To denote the quantity completely, what additional information must be given besides the number 40?

      How would you *interpret* the number 40 in this instance? (*Note:* A name is *not* an interpretation. Your response should be in terms of centimeters and seconds.)

      Use your interpretation (not algebra) to find the distance the train moves in 2.5 s.

   2.   Is there a name that is commonly given to the quantity represented by the number 0.025? (0.025 = 1.5/60) If so, what is the name?

      How would you *interpret* the number 0.025?

      Use your interpretation (not algebra) to find the time it takes the train to move 90 cm.

II.   Motion with varying speed

A.   In the space below, sketch a possible ticker tape resulting from motion with varying speed and write a description of the motion.

How can you tell from your diagram that the motion has varying speed?

B.   Together with your classmates, take your ticker tapes and arrange yourselves in a line, ranked according to the speed of the tapes. Discuss the following questions as a class.

Compare your segment of ticker tape to neighboring tape segments. What do you observe?

Compare the smallest and largest speeds. What do you observe?

C.   Based on your observations of your tape and the tapes of other members of your class, answer the following questions.

Is each segment of tape a part of a motion with constant or varying speed?

Did your examination of a single, small section of the tape reveal whether the motion that generated the tape had constant or varying speed?

D. Review your earlier interpretation of the speed for your piece of tape. (See section I.) Is that interpretation valid for the entire motion that generated the tape?

Based on the speed for your piece of tape, could you successfully predict how far the object would move in: $1/60^{th}$ s? $2/60^{th}$ s? 1 s?

How can you modify the interpretation of the speed so that it applies even to motion with varying speed?

What name do we give to a speed that is interpreted in this way?

E. Suppose you selected two widely separated dots on the assembled ticker tape. What would you call the number you would obtain if you divided the distance between the dots by the time it took the object to move between the dots?

How would you interpret this number?

Interpret - don't like.

Tutorials in Introductory Physics
McDermott, Shaffer, & P.E.G., U.Wash.

©Prentice Hall
Preliminary Edition, 1998

In this tutorial, you will use a motion detector to graph your motion and to investigate how motion can be described in terms of position, velocity, and acceleration. See your instructor for instructions on using the equipment.

General tips

When using a motion detector:

• Stay in line with the detector and do not swing your arms. For best results, take off bulky sweaters or other loose-fitting clothing. You may find it helpful to hold a large board in front of you in order to present a larger target for the detector.

• Do not stand closer than 0.5 meter or farther than 4 meters from the detector.

• It is difficult to obtain good *a* versus *t* graphs with the motion detector. Discuss any questions about your *a* versus *t* graphs with an instructor.

Instructions

In each of the following problems, you will be given one of the following descriptions of a motion:

• a written description, or

• an *x* versus *t, v* versus *t,* or *a* versus *t* graph.

Predict the other three descriptions of the motion, then use the motion detector to check your answers. Check your predictions one-by-one, instead of checking a whole page at once. In addition, answer the questions posed at the bottom of each page.

Note: So that your graphs emphasize important features, draw them in an idealized form rather than showing many small wiggles.

Example: The problem below has been worked as an example. Use the motion detector to verify the answers.

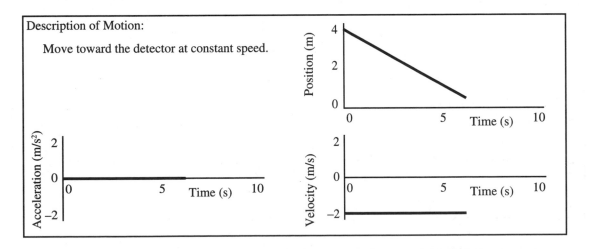

The computer program assumes a particular coordinate system. Describe this coordinate system.

Tutorials in Introductory Physics
McDermott, Shaffer, & P.E.G., U.Wash.

©Prentice Hall
Preliminary Edition, 1998

A.

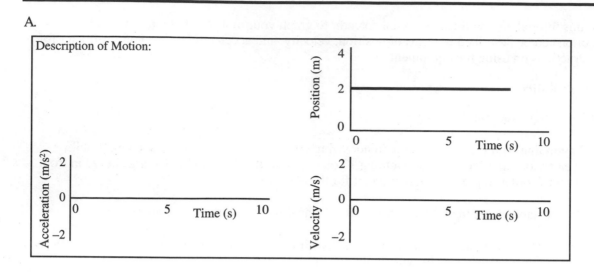

B.

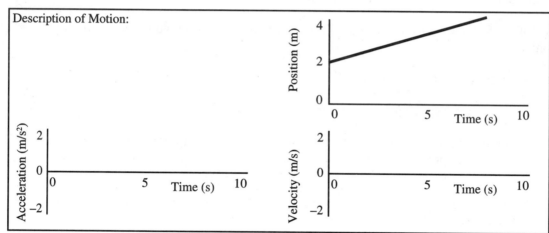

C.

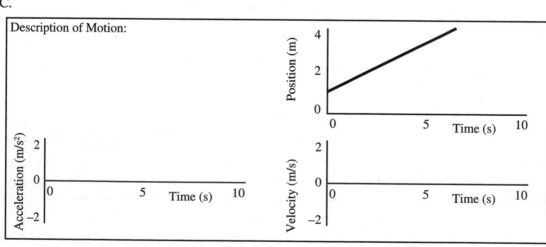

How are the motions of B and C similar? How do they differ? How are the graphs similar? How do they differ?

Tutorials in Introductory Physics
McDermott, Shaffer, & P.E.G., U.Wash.

©Prentice Hall
Preliminary Edition, 1998

D.

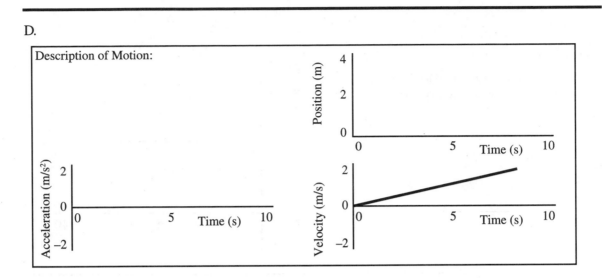

Description of Motion:

E.

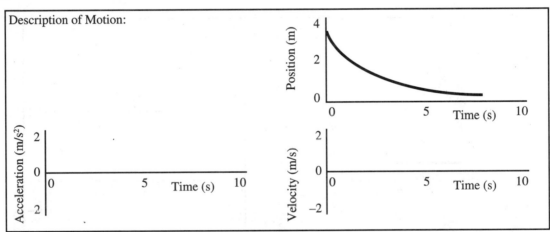

Description of Motion:

F.

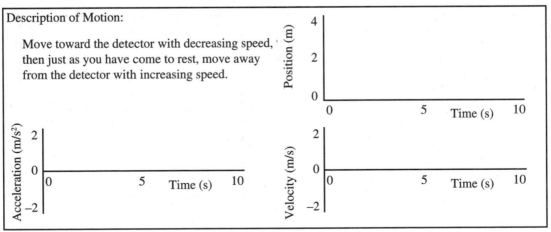

Description of Motion:

Move toward the detector with decreasing speed,
then just as you have come to rest, move away
from the detector with increasing speed.

How do the acceleration graphs for D, E, and F compare? Is it possible to: have a positive
acceleration and slow down? have a negative acceleration and speed up?

G.

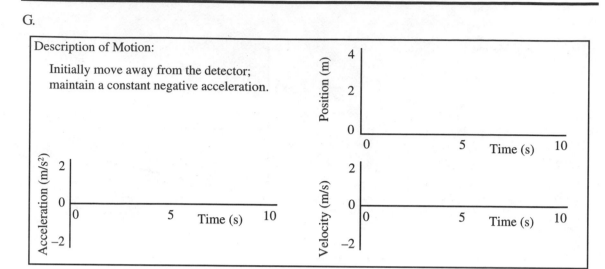

Description of Motion:

Initially move away from the detector;
maintain a constant negative acceleration.

H.

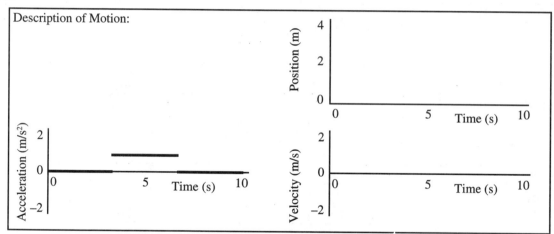

Description of Motion:

I.

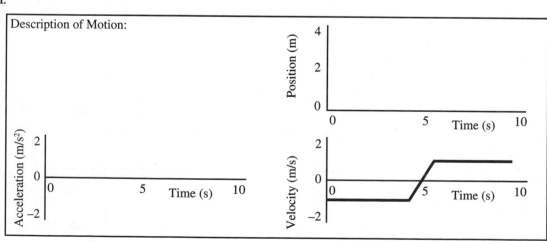

Description of Motion:

The term *decelerate* is often used to indicate that an object is slowing down. Does this term indicate the sign of the acceleration?

I. Motion with decreasing speed

The diagram below represents a strobe photograph of a ball as it rolls *up* a track. (In a strobe photograph, the position of an object is shown at instants separated by *equal time intervals*.)

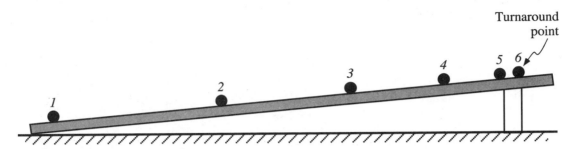

Turnaround
point

Draw vectors on your diagram that represent the instantaneous velocity of the ball at each of the labeled locations. If the velocity is zero at any point, indicate that explicitly. Explain why you drew the vectors as you did.

We will call diagrams like the one you drew above *velocity diagrams*. Unless otherwise specified, a velocity diagram shows both the location and the velocity of an object at instants in time that are separated by equal time intervals.

A. In the space at right, compare the velocities at points *1* and *2* by sketching the vectors that represent those velocities. Draw the vectors side-by-side and label them $\vec{v}_1$ and $\vec{v}_2$, respectively.

Sketch $\vec{v}_1$, $\vec{v}_2$, and $\Delta\vec{v}$

Draw the vector that must be *added* to the velocity at the earlier time to equal the velocity at the later time. Label this vector $\Delta\vec{v}$.

Why is the name *change in velocity* appropriate for this vector?

How does the direction of the change in velocity vector compare to the direction of the velocity vectors?

Would your answer change if you were to select two *different* consecutive points (*e.g.,* points *3* and *4*) while the ball was slowing down? Explain.

this
seems
difficult

What answer?

Tutorials in Introductory Physics
McDermott, Shaffer, & P.E.G., U.Wash.

©Prentice Hall
Preliminary Edition, 1998

How would the magnitude of the change in velocity vector between points *1* and *2* compare to the magnitude of the change in velocity vector between two *different* consecutive points *(e.g., points 3 and 4)?* Explain. (You may find it useful to refer to the v versus t graph for the motion of the ball as it rolls up the track.)

(*Note:* The positive direction has been chosen to be up the track.)

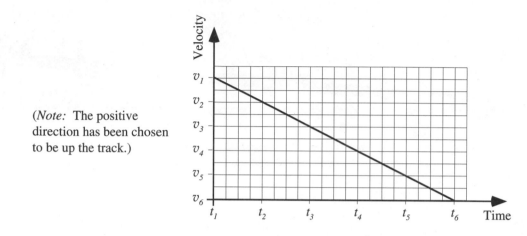

B. Consider the change in velocity vector between two points on the velocity diagram that are not consecutive, *e.g.*, points *1* and *4.*

Is the direction of the change in velocity vector different than it was for consecutive points? Explain.

Is the length of the change in velocity vector different than it was for consecutive points? If so, how many times larger or smaller is it than the corresponding vector for consecutive points? Explain.

C. Use the definition of acceleration to draw a vector in the space at right that represents the acceleration of the ball between points *1* and *2.*

Sketch acceleration vector

How is the direction of the acceleration vector related to the direction of the change in velocity vector? Explain.

D. Does the acceleration change as the ball rolls up the track? Would the acceleration vector you obtain differ if you were to choose (1) two different successive points on your diagram or (2) two points that are not consecutive? Explain.

©Prentice Hall
Preliminary Edition, 1998

E. Generalize your results to answer the following question:

What is the relationship between the direction of the acceleration and the direction of the velocity for an object that is slowing down? Explain.

Describe the direction of the acceleration of a ball that is rolling up an incline.

II. Motion with increasing speed

The diagram below represents a strobe photograph of a ball as it rolls *down* the track.

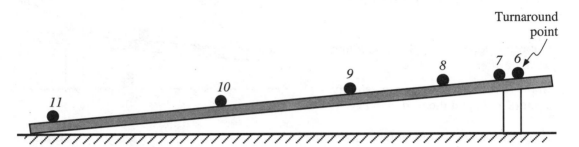

A. Choose two successive points. In the space at right, sketch the velocity vectors corresponding to those points. Draw the vectors side-by-side and label them $\vec{v}_i$ and $\vec{v}_f$, respectively.

Sketch $\vec{v}_i$, $\vec{v}_f$, and $\Delta\vec{v}$

Determine the vector that must be added to the velocity at the earlier time to equal the velocity at the later time. Is the name *change in velocity* appropriate for this vector?

How does the direction of the change in velocity vector compare to the direction of the velocity vectors in this case?

Would your answer change if you were to select two *different* points during the time that the ball was speeding up? Explain.

B. In the space at right, draw a vector to represent the acceleration of the ball between the points chosen above.

Sketch acceleration vector

How is the direction of the change in velocity vector related to the direction of the acceleration vector? Explain.

C. Generalize the results above to answer the following question:

What is the relationship between the direction of the acceleration and the direction of the velocity for an object that is speeding up? Explain.

Describe the direction of the acceleration of a ball that is rolling down an incline.

III. Motion that includes a turnaround

Complete the velocity diagram at right for the portion of the motion that includes the turnaround.

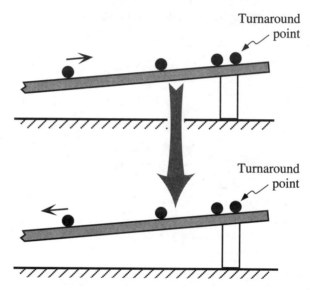

Turnaround point

Turnaround point

A. Choose a point *before* the turnaround and another *after*.

In the space below, draw the velocity vectors and label them $\vec{v}_i$ and $\vec{v}_f$.

Draw the vector that must be added to the velocity at the earlier time to obtain the velocity at the later time.

Is the name *change in velocity* that you used in sections I and II also appropriate for this vector?

B. Suppose that you had chosen the turnaround as one of your points.

What is the velocity at the turnaround point?

Would this choice affect the direction of the change in velocity vector? Explain why or why not.

Sketch $\vec{v}_i$, $\vec{v}_f$, and $\vec{\Delta v}$

C. In the space at right, draw a vector that represents the acceleration of the ball between the points you chose in part A above.

Sketch acceleration vector

Compare the direction of the acceleration of the ball at the turnaround point to that of the ball as it rolls: (1) *up* the track and (2) *down* the track.

Tutorials in Introductory Physics
McDermott, Shaffer, & P.E.G., U.Wash.

©Prentice Hall
Preliminary Edition, 1998

I.   Velocity

An object is moving around an oval track. Sketch the trajectory of the object on a large sheet of paper. (Make your diagram *large*.)

A. Choose a point to serve as an origin for your coordinate system. Label that point *O* (for origin). Select two locations of the object that are about one-eighth of the oval apart and label them *A* and *B*.

1. Draw the position vectors for each of the two locations *A* and *B* and draw the vector that represents the displacement from *A* to *B*.

> Copy your group's drawing in this space after discussion.

2. Describe how to use the displacement vector to determine the average velocity of the object between *A* and *B*. Draw a vector to represent the average velocity.

3. Choose a point between points *A* and *B*, and label that point *B´*.

    As point *B´* is chosen to lie closer and closer to point *A*, how does the direction of the average velocity over the interval *AB´* compare to the direction of the average velocity over the interval *AB*? more + more different

4. Describe the direction of the instantaneous velocity of the object at point *A*.

    How would you characterize the direction of the instantaneous velocity at *any* point on the trajectory?

    Does your answer depend on whether the object is speeding up, slowing down, or moving with constant speed? Explain.

B. If you were to choose a different origin for the coordinate system, which of the vectors that you have drawn in part A would change and which would not change?

Tutorials in Introductory Physics
McDermott, Shaffer, & P.E.G., U.Wash.

©Prentice Hall
Preliminary Edition, 1998

II.   Acceleration for motion with constant speed

Suppose that the object in section I is moving around the track at *constant speed*. Draw vectors to represent the velocity at two points on the track that are relatively close together. (Draw your vectors *large*.) Label the two points *C* and *D*.

A.   On a *separate* part of your paper, copy the velocity vectors $\vec{v}_C$ and $\vec{v}_D$. From these vectors, determine the *change in velocity vector*, $\Delta\vec{v}$.

   1.   How does the angle formed by the "head" of $\vec{v}_C$ and the "tail" of $\Delta\vec{v}$ compare to 90°? ("Compare" in this case means "is it *less than, greater than,* or *equal to* 90°?")

   As point *D* is chosen to lie closer and closer to point *C*, does the above angle *increase, decrease,* or *remain the same?* Explain how you can tell.

   Does the above angle approach a *limiting* value? If so, what is its limiting value?

   2.   Describe how to use the change in velocity vector to determine the average acceleration of the object between *C* and *D*. Draw a vector to represent the average acceleration between points *C* and *D*.

   What happens to the magnitude of $\Delta\vec{v}$ as point *D* is chosen to lie closer and closer to point *C*? Does the acceleration change in the same way? Explain.

   Consider the direction of the acceleration at point *C*. How does the angle between the acceleration vector and the velocity vector compare to 90°? (*Note:* Conventionally, the angle between two vectors is defined as the angle formed when they are placed "tail-to-tail.")

Tutorials in Introductory Physics
McDermott, Shaffer, & P.E.G., U.Wash.

©Prentice Hall
Preliminary Edition, 1998

B. Suppose you were to choose a new point on the trajectory where the *curvature is different* from that at point *C*.

How would the magnitude of the acceleration at the new point compare to the magnitude of the acceleration at point *C*? Explain.

Describe the direction of the acceleration at the new point.

⇨   Check your reasoning with a tutorial instructor before proceeding.   *Practical?*

III. Acceleration for motion with changing speed

Suppose that the object is *speeding up* as it moves around the track. Draw vectors to represent the velocity at two points on the track that are relatively close together. (Draw your vectors *large*.) Label the two points *E* and *F*.

A. On a *separate* part of your paper, copy the velocity vectors $\vec{v}_E$ and $\vec{v}_F$. From these vectors, determine the change in velocity vector, $\Delta\vec{v}$.

1. How does the angle formed by the head of $\vec{v}_E$ and the tail of $\Delta\vec{v}$ compare to 90°?

Consider this angle as point *F* is chosen to lie closer and closer to point *E*.

What value or range of values is possible for this angle for an object that is speeding up? Explain.   *90 to 180*

What happens to the magnitude of $\Delta\vec{v}$ as point *F* is chosen to lie closer and closer to point *E*?   *gets smaller*

2. Describe how you would use the change in velocity vector to determine the acceleration of the object at point *E*.

Consider the direction of the acceleration at point *E*. How does the angle between the acceleration vector and the velocity vector (placed "tail-to-tail") compare to 90°?

B. Suppose the object started *from rest* at point *E* and moved towards point *F* with increasing speed. How would you find the acceleration at point *E?*

  Describe the direction of the acceleration of an object speeding up from rest.

C. At several points on each of the diagrams below, draw a vector that represents the acceleration of the object.

Acceleration vectors for constant speed   Acceleration vectors for speeding up

Top view diagram   Top view diagram

Characterize the direction of the acceleration at each point on the trajectory for each case.

  Is the acceleration directed toward the "center" of the oval at every point on the trajectory for either of these cases?

  How could you characterize the direction of the acceleration if the trajectory were:

  • peanut-shaped?

Constant speed   Speeding up

  • circular?

Newton's laws

Tutorials in Introductory Physics
McDermott, Shaffer, & P.E.G., U.Wash.

©Prentice Hall
Preliminary Edition, 1998

I.   Free-body diagrams

Two people are attempting to move a large block. The block, however, does not move. Chris is pushing on the block. Pam is pulling on a rope attached to the block.

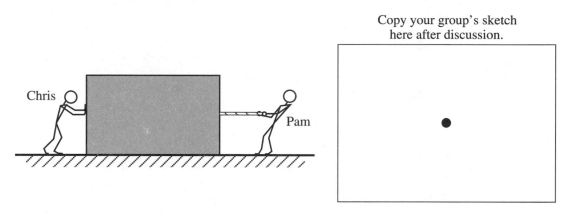

Copy your group's sketch
here after discussion.

A. Draw a large dot on your large sheet of paper to represent the block. Draw vectors with their "tails" on the dot to show the forces exerted *on* the block. Label each vector and write a brief description of that force next to the vector.

In Newtonian physics, all forces arise from an interaction between *two* objects. Forces are specified by identifying the object *on which* the force is exerted and the object *that is exerting* the force. For example, in the situation above, a gravitational force is exerted *on* the block *by* the earth.

B. Describe the remaining forces you have indicated above in a similar fashion.

The diagram you have drawn is called a *free-body diagram*. A free-body diagram should show only the forces exerted *on* the object or system of interest, that is, in this case, *on the block*. Check your free-body diagram and, if necessary, modify it accordingly.

Sometimes a free-body diagram involves a simplified sketch of the object rather than the dot. (Your instructor will indicate the convention you are to use.) Regardless of which form is used, a proper free-body diagram should *not* have anything on it except a representation of the object and the (labeled) forces exerted on that object. A free-body diagram *never* includes (1) forces exerted by the object of interest on other objects or (2) sketches of other objects that exert forces on the object of interest.

Tutorials in Introductory Physics
McDermott, Shaffer, & P.E.G., U.Wash.

©Prentice Hall
Preliminary Edition, 1998

C. All forces arise from interactions between objects, but the interactions can take different forms.

        Which of the forces exerted on the block require *direct contact* between the block and the object exerting the force?

        Which of the forces exerted on the block *do not* arise from direct contact between the block and the object exerting the force?

We will call forces that depend on contact between two objects *contact forces*. We will call forces that do not arise from contact between two objects *non-contact forces*.

D. There are many different types of forces, including: friction ($\vec{f}$), tension ($\vec{T}$), magnetic forces ($\vec{F}^{mag}$), normal forces ($\vec{N}$), and the gravitational force ($\vec{W}$, for weight). Categorize these forces according to whether they are contact or non-contact forces.

<u>Contact forces</u>   <u>Non-contact forces</u>

E. Consider the following discussion between two students.

    Student 1:   *"I think the free-body diagram for the block should have a force by Chris, a force by the rope, and a force by Pam."*

    Student 2:   *"I don't think the diagram should show a force by Pam. People can't exert forces on blocks without touching them."*

With which student, if either, do you agree? Explain your reasoning.

It is often useful to label forces in a way that makes clear (1) the type of force, (2) the object on which the force is exerted, and (3) the object exerting the force. For example, the gravitational force exerted *on* the block *by* the earth might be labeled $\vec{W}_{BE}$. Your instructor will indicate the notation that you are to use.

F. Label each of the forces on your free-body diagram in part A in the manner described above.

⇨ Do not proceed until a tutorial instructor has checked your free-body diagram.

Tutorials in Introductory Physics
McDermott, Shaffer, & P.E.G., U.Wash.

©Prentice Hall
Preliminary Edition, 1998

II. Forces

A. Sketch a free-body diagram for a book at rest on a level table.  Book
(*Remember:* A proper free-body diagram should not have anything on
it except a representation of the book and the forces exerted *on* the
book.)

Make sure the label for each force indicates:

• the type of force (gravitational, frictional, *etc.*),
• the object on which the force is exerted, and
• the object exerting the force.

1. What evidence do you have for the existence of each of the forces on your diagram?

2. What observation can you make that allows you to determine the relative magnitudes of
the forces acting on the book?

How did you show the relative magnitudes of the forces on your diagram?

B. A second book of greater mass is placed on top of the first.

Upper book
Lower book

Sketch a free-body diagram for each of the books in the space
below. Label all the forces as in part A.

Free-body diagram for upper book	Free-body diagram for lower book

Specify which of the forces are contact forces and which are non-contact.

---

1. Examine all the forces on the two free-body diagrams you just drew. Explain why a force that appears on one diagram *should not* appear on the other diagram.

2. What *type* of force does the upper book exert on the lower book (*e.g.,* frictional, gravitational)?

   Why would it be *incorrect* to say that the weight of the upper book acts on the lower book?

3. What observation can you make that allows you to determine the relative magnitudes of the forces on the *upper* book?

4. Are there any forces acting on the *lower* book that have the same magnitude as a force acting on the *upper* book? Explain.

C. Compare the free-body diagram for the lower book to the free-body diagram for the same book in part A (*i.e.,* before the upper book was added).

   Which of the forces changed when the upper book was added and which remained the same?

As discussed earlier, we think of each force acting on an object as being exerted by another object. The first object exerts a force of equal magnitude and opposite direction on the second object. The two forces together are called an *action-reaction* or *Newton's third law* force pair.

D. Which, if any, Newton's third law force pairs are shown in the diagrams you have drawn? On which object does each of the forces in the pair act?

   Identify any third law force pairs on your diagrams by placing a small "✕" through each member of the pair. For example, if you have two sets of third law force pairs shown on your diagrams, mark *each* member of the first pair as ➝✕▶, and each member of the second pair as ✕✕▶.

III. Supplement: Contact and non-contact forces

A. A magnet is supported by another magnet as shown at right.

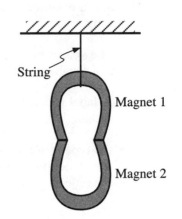

String

Magnet 1

Magnet 2

1. Draw a free-body diagram for magnet 2. The label for each of the forces on your diagram should indicate:

   • the type of force (*e.g.,* gravitational, normal),
   • the object on which the force is exerted, and
   • the object exerting the force.

2. Suppose that the magnets were replaced by stronger magnets of the same mass.

   If this changes the free-body diagram for magnet 2, sketch the new free-body diagram and describe how the diagram changes. (Label the forces as you did in part 1 above.) If the free-body diagram for magnet 2 does not change, explain why it does not.

3. Can a magnet exert a non-contact force on another object?

   Can a magnet exert a contact force on another object?

   Describe how you can use a magnet to exert *both* a contact force and a non-contact force on another magnet.

4. To ensure that you have accounted for all the forces acting on magnet 2 in parts 1 and 2:

   List all the non-contact forces acting on magnet 2.

   List all the contact forces acting on magnet 2. (*Hint:* Which objects are in *contact* with magnet 2?)

©Prentice Hall
Preliminary Edition, 1998

B. An iron rod is held up by a magnet as shown. The magnet is held up by a string.

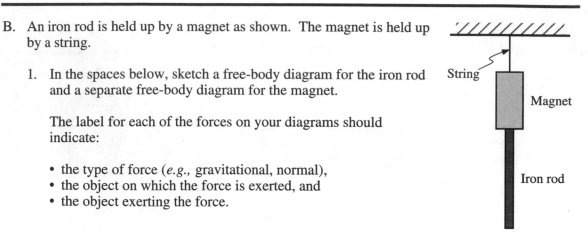

1. In the spaces below, sketch a free-body diagram for the iron rod and a separate free-body diagram for the magnet.

    The label for each of the forces on your diagrams should indicate:

    • the type of force (*e.g.,* gravitational, normal),
    • the object on which the force is exerted, and
    • the object exerting the force.

Free-body diagam for iron rod	Free-body diagam for magnet
(*Hint:* There should be three forces.)	(*Hint:* There should be four forces.)

2. For each of the forces shown in your diagram for the iron rod, identify the corresponding force that completes the Newton's third law (or action-reaction) force pair.

3. How would your diagram for the iron rod change if the magnet were replaced with a stronger magnet? Which forces would change (in type or in magnitude)? Which forces would remain the same?

I.   Interacting objects: constant speed

Three identical bricks are pushed across a table at *constant speed* as shown. The hand pushes
horizontally. (*Note:* There is friction between the bricks and the table.)

Call the stack of two bricks system A and the single brick,
system B.

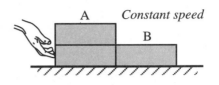

A. Describe the motions of systems A and B.

B. Compare the *net force* (magnitude and direction) on system A to that on system B. Explain
    how you arrived at your comparison.

C. Draw separate free-body diagrams for system A and system B. Label each of the forces in
    your diagrams by identifying: the type of force, the object on which the force is exerted, and
    the object exerting the force.

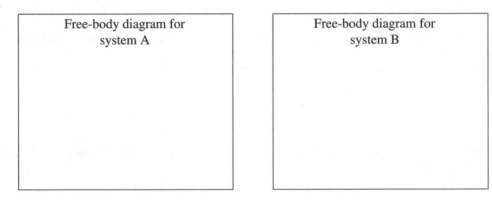

D. Is the magnitude of the force exerted on system A by system B *greater than, less than,* or
    equal to the magnitude of the force exerted on system B by system A? Explain.

    Would your answer change if the hand were pushing system B to the left instead of
    pushing system A to the right? If so, how? If not, why not?

E. Identify any *Newton's third law* (or *action-reaction*) force pairs you have drawn using the
    convention introduced in the *Forces* tutorial, that is, by placing a small "✕" through each
    member of the pair.

    What criteria did you use to identify the force pair(s)?

    Is your answer to part D consistent with your identification of Newton's third law (or
    action-reaction) force pairs? If so, explain how it is consistent. If not, resolve the
    inconsistency.

F. Rank the magnitudes of all the *horizontal* forces that you identified on your free-body diagrams in part C. (*Hint:* Recall that the bricks are pushed so that they move at constant speed.)

Did you apply Newton's second law in comparing the magnitudes of the horizontal forces? If so, how?

Did you apply Newton's third law in comparing the magnitudes of the horizontal forces? If so, how?

What information besides Newton's laws did you need to apply in comparing the magnitudes of the horizontal forces?

G. Suppose the mass of each brick is 2.5 kg, the coefficient of kinetic friction between the bricks and the table is 0.2, and the bricks are moving at a constant speed of 0.50 m/s.

Determine the magnitude of each of the forces that you drew on your free-body diagrams in part C. (Use the approximation $g = 10$ m/s^2.)

Would your answers change if the bricks were moving half as fast? If so how? If not, why not?

⇨ Discuss your answers with a tutorial instructor before continuing.

Tutorials in Introductory Physics
McDermott, Shaffer, & P.E.G., U.Wash.

©Prentice Hall
Preliminary Edition, 1998

II.   Interacting objects: varying speed

Suppose the bricks were pushed by the hand with the same force as in section I; however, the coefficient of kinetic friction between the bricks and the table is *less than* that in section I.

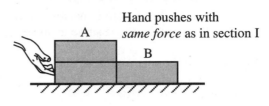

A.   Describe the motions of systems A and B.

B.   Compare the *net force* (magnitude and direction) on system A to that on system B. Explain.

C.   Draw and label separate free-body diagrams for systems A and B.

Free-body diagram for system A

Free-body diagram for system B

D.   Consider the following discussion between two students.

   Student 1:   *"System A and system B are pushed by the same force as before, so they will have the same motion as in section I."*

   Student 2:   *"I disagree. I think that they are speeding up since friction is less. So now system A is pushing on system B with a greater force than system B is pushing on system A."*

   With which student, if either, do you agree? Explain your reasoning.

E.   Rank the magnitudes of all the *horizontal* forces that appear on your free-body diagrams in part C. Explain your reasoning. (Describe explicitly how you used Newton's second and third laws to compare the magnitudes of the forces.)

      Is it possible to *completely* rank the horizontal forces in this case?

III. System of interacting objects

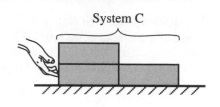

Let C represent the system consisting of all three bricks. The motion of the blocks is the same as in section II.

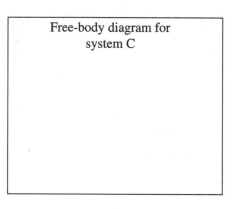

A. Compare the magnitude of the *net force* on system C to the magnitudes of the *net forces* on systems A and B. Explain.

B. Draw and label a free-body diagram for system C.

Compare the forces that appear on your free-body diagram for system C to those that appear on your diagrams for systems A and B in section II.

For each of the forces that appear on your diagram for system C, list the corresponding force (or forces) on your diagrams for systems A and B.

> Free-body diagram for
> system C

Are there any forces on your diagrams for systems A and B that you did not list? If so, what characteristic do these forces have in common that none of the others share?

Why is it not necessary to consider these forces in determining the motion of system C?

Note that such forces are sometimes called *internal forces,* to be distinguished from *external forces.*

IV. Application of Newton's laws

At right is a free-body diagram for a cart. All forces have been drawn to scale.

In the space below, sketch the cart, rope, *etc.,* as they would appear in the laboratory.

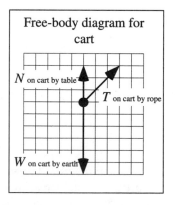

Free-body diagram for cart

What can you say about the motion of the cart based on the free-body diagram? For example, could the cart be: moving to the left? moving to the right? stationary? Explain whether each case is possible and, if so, describe the motion of the cart.

©Prentice Hall
Preliminary Edition, 1998

I. Blocks connected by a rope

Two blocks, A and B, are tied together with a rope of mass M. Block B is being pushed with a constant horizontal force as shown at right. Assume that there is no friction between the blocks and the table and that the blocks have already been moving for a while at the instant shown.

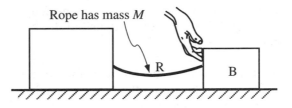

Rope has mass M

R

B

A. Describe the motions of block A, block B, and the rope.

B. Draw a separate free-body diagram for each block and for the rope. Clearly label the forces.

Copy your free-body diagrams below after discussion.

Free-body diagram for block A	Free-body diagram for rope	Free-body diagram for block B

C. Identify all the *Newton's third law (action-reaction)* force pairs in your diagrams by placing a small "×" through each member of the pair (*i.e.*, mark each member of the first pair as —✕▶, each member of the second pair as ✕✕▶, and so on).

D. Compare the magnitudes of the *horizontal components* of all forces on your diagrams. If any of the forces have the same magnitude, state that explicitly. Explain the reasoning you used to arrive at this comparison.

E. Consider the horizontal components of the forces exerted *on the rope* by blocks A and B. Is your answer above for the relative magnitude of these components consistent with your knowledge of the net force on the rope?

Tutorials in Introductory Physics
McDermott, Shaffer, & P.E.G., U.Wash.

©Prentice Hall
Preliminary Edition, 1998

II. Blocks connected by a very light string

The blocks in section I are now connected with a very light, flexible, and inextensible string of mass m $(m < M)$.

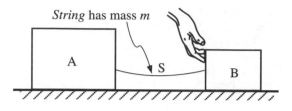

String has mass m

A

S

B

A. If the motion of the blocks is the same as in section I, how does the net force on the *string* compare to the net force on the *rope?*

Determine whether the net force on the following objects is *greater than, less than,* or *equal to* the net force on them in section I: block A, block B, and the system composed of the blocks and the connecting rope or string. Explain.

Compare the horizontal components of the following pairs of forces:

- the force on the string by block A and the force on the rope by block A. Explain.

- the force on the string by block B and the force on the rope by block B. Explain.

B. Suppose the mass of the string that connects blocks A and B becomes smaller and smaller, but the motion remains the same as in section I. What happens to:

- the magnitude of the net force on that connecting string?

- the magnitudes of the forces exerted on that connecting string by blocks A and B?

C. A string exerts a force on each of the two objects to which it is attached. For a massless string, the magnitudes of both forces are often referred to as "the tension in the string."

Justify the use of this terminology, in which a *single value* is assumed for the magnitudes of both forces.

D. If you know that the net force on a massless string is zero, what can you infer about its motion?

Is it possible to exert a non-zero force on a massless string? Is it possible for a massless string to have a non-zero *net* force? Explain.

⇨ Discuss your answers above with a tutorial instructor before continuing.

III. The Atwood's machine

The Atwood's machine at right consists of two identical objects connected by a massless string that runs over an ideal pulley. Object B is initially held so that it is above object A and so that neither object can move.

A. Predict the subsequent motions of objects A and B after they are released. Explain the basis for your description. Do not use algebra.

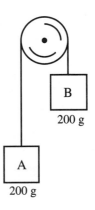

B. Draw separate free-body diagrams for objects A and B. Are your free-body diagrams consistent with your prediction of the motion of the objects?

Object B is replaced by object C, of greater mass. Object C is initially held so that it is higher than object A and so that neither object can move.

C. *Predict:*

• what will happen to object C when it is released.

• how the motion of object C will compare to the motion of object A after they are released.

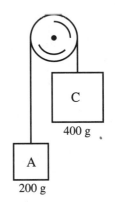

Explain the basis for your predictions. Do not use algebra.

Tutorials in Introductory Physics
McDermott, Shaffer, & P.E.G., U.Wash.

©Prentice Hall
Preliminary Edition, 1998

D. Draw and label separate free-body diagrams for objects A and C *after* they are released. Indicate the relative magnitudes of the forces by the relative lengths of the force vectors.

Are the predictions you made in part C consistent with your free-body diagrams for objects A and C? If so, explain why they are consistent. If not, then resolve the inconsistency.

E. The weight of a 200 g mass has magnitude $(0.2 \text{ kg})(9.8 \text{ m/s}^2) \approx 2 \text{ N}$. Similarly, the weight of a 400 g mass is approximately 4 N in magnitude.

1. How does the force exerted on object A by the string compare to these two weights?

2. How does the force exerted on object C by the string compare to these two weights?

Explain your answers.

3. How does the net force on object A compare to the net force on object C? Explain.

F. Consider the following statement about the Atwood's machine made by a student.

"All strings can do is transmit forces from other objects. That means that the string in the Atwood's machine just transmits the weight of one block to the other."

Do you agree with this student? Explain your reasoning.

Tutorials in Introductory Physics
McDermott, Shaffer, & P.E.G., U.Wash.

©Prentice Hall
Preliminary Edition, 1998

Energy and momentum

I. Changes in kinetic energy

A glider moves on a level air track while a hand applies a constant horizontal force ($\vec{F}_{GH}$). Friction is negligible. The glider starts from rest at point A. The hand continues to push with the same force along the entire length of track.

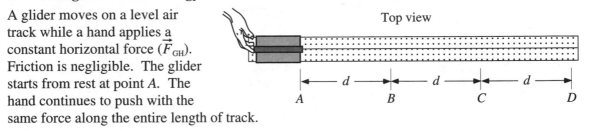

A. Describe the motion of the glider. Explain how you determined your answer.

Suppose that the time interval for the glider to move from point A to point B is Δt_1. Is the time taken by the glider to move from point B to point C *greater than, less than,* or *equal to* Δt_1? Explain.

Suppose the speed of the glider increases by Δv_1 from point A to point B. Is the increase in speed from point B to point C *greater than, less than,* or *equal to* Δv_1? Explain how the definition of acceleration can be used to determine your answer.

B. For one-dimensional motion with constant acceleration, the final speed of an object is related to its displacement Δs by the formula $v_f^2 = v_i^2 + 2a\Delta s$.

Use this formula to write expressions for the speed of the glider at points B and C in terms of the acceleration of the glider, a, and the distance d.

Use your answer to obtain expressions for the *change in speed* of the glider between points A and B (Δv_{AB}) and between points B and C (Δv_{BC}).

Is Δv_{BC} *greater than, less than,* or *equal to* Δv_{AB}?

Is your answer consistent with your answer to part A? If not, resolve the inconsistency.

C. The *kinetic energy* of an object is defined to be $\frac{1}{2}mv^2$.

Use the formula given in part B to derive an equation for the *final* kinetic energy of an object in terms of the net force on it, the distance Δs that it has moved, and its *initial* kinetic energy.

Tutorials in Introductory Physics
McDermott, Shaffer, & P.E.G., U.Wash.

©Prentice Hall
Preliminary Edition, 1998

Now express the *change in kinetic energy* (ΔKE) of an object in terms of the net force on the object and the distance Δs that it has moved.

The relationship you have just derived applies to any rigid object moving in one dimension in the direction of a constant net force. State this relationship *in words*.

II. Work

Suppose now that instead of pushing horizontally, the hand pushes with the same magnitude force as before, F_{GH}, but at an angle θ down and to the right. The displacement of the glider is $\Delta \vec{s}$.

A. Draw a free-body diagram for the glider.

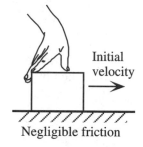

Initial velocity

Negligible friction

B. Write an equation showing the relationship between the magnitude of the net force on the glider and the magnitude of the force on the glider by the hand.

C. Use your answer and the result from section I to write an equation giving the change in kinetic energy of the glider in terms of θ and the magnitudes F_{GH} and $|\Delta \vec{s}|$.

Rewrite this equation in terms of a product between the force vector, $\vec{F}_{GH}$, and the displacement vector, $\Delta \vec{s}$.

Express *in words* the product in two different ways: one involving the component of the force along the direction of displacement; another involving the component of the displacement along the direction of the force.

How could the hand push on the glider to make this product negative?

We refer to the quantity $\vec{F} \bullet \Delta \vec{s}$ as the *work* done on a rigid object by the agent exerting the force $\vec{F}$. For example, the quantity $\vec{F}_{GH} \bullet \Delta \vec{s}$ is called "the work done on the glider by the hand." In part C, you derived an equation relating the change in the kinetic energy of an object to the net force on the object. This equation is known as the *work-energy theorem*.

D. State the work-energy theorem *in words*.

E. In each of the following cases, an object undergoes a displacement $\Delta\vec{s}$ while constant forces
    are exerted on it. In each case, only one force is shown. Determine whether the work done
    on the object by the agent exerting the indicated forces is *positive, negative,* or *zero.*

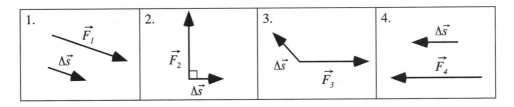

III. Applications of the work-energy theorem

A. An elevator of weight W is initially at rest on the first floor of a building. It moves upward
    and passes the second floor with speed v_o. It continues upward and finally stops at the fourth
    floor. The distance between adjacent floors is H.

1. In the space below, draw a free-body diagram for the elevator
    immediately after it leaves the first floor. Make sure to draw all
    forces with correct relative magnitudes.

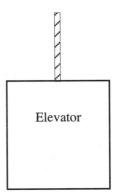

Elevator

2. Is the change in kinetic energy of the elevator as it travels from the
    first floor to the *second* floor *greater than, less than,* or *equal to*
    zero? Explain.

    Based on your answer above, is the net work done on the elevator as it travels from the
    first floor to the second floor *greater than, less than,* or *equal to* zero? Explain.

        Is your answer for the sign of the net work consistent with your free-body diagram
        for the elevator? If so, explain how. If not, resolve the inconsistency.

3. Is the change in kinetic energy for the *entire trip* (from the first floor to the fourth floor)
    greater than, less than, or *equal to* zero? Explain.

    Based on your answer above, is the net work done on the elevator during the entire trip
    greater than, less than, or *equal to* zero? Explain.

4. Write an expression for the work done on the elevator by the cable over the entire trip in terms of the given quantities only. Indicate whether this work is *positive* or *negative*.

Does your result imply that the force on the elevator by the cable is always equal in magnitude to the weight of the elevator? Explain.

B. Two identical blocks are released from rest from the same vertical height on frictionless inclines. The final speed of each block is denoted by v_{Af} and v_{Bf}. The distance that block A travels, l_1, is smaller than the distance that block B travels, l_2.

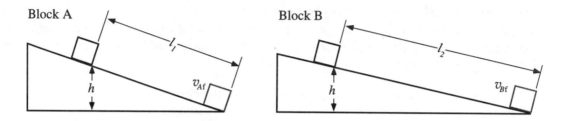

1. In the space below, draw separate free-body diagrams for blocks A and B at an instant during their motion.

Determine whether the work done by each force identified on your free-body diagrams is *positive, negative,* or *zero.* Explain your reasoning.

2. Use the definition of work to compare the net work done on blocks A and B. Explain how you arrived at your comparison. (*Hint:* Recall that the dot product of two vectors can be thought of in two ways.)

Based on your answer, is the final speed of A, v_{Af}, *greater than, less than,* or *equal to* the final speed of B, v_{Bf}?

Tutorials in Introductory Physics
McDermott, Shaffer, & P.E.G., U.Wash.

©Prentice Hall
Preliminary Edition, 1998

I.   Kinetic energy and momentum

Two carts, A and B, are initially at rest on a horizontal frictionless table. The same constant force $\vec{F}$ is exerted on each cart as it travels between two marks on the table (see diagram). The carts are then allowed to glide freely. The mass of cart A is *less* than the mass of cart B.

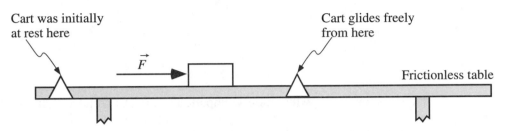

A.   Three students discuss the final momentum and kinetic energy of each cart.

Student 1:   *"Since the same force is exerted on both carts, the cart with the smaller mass will move quickly, while the cart with the larger mass will move slowly. The momentum of each cart is equal to its mass times its velocity."*

Student 2:   *"This must mean that the speed compensates for the mass and the two carts have equal final momenta."*

Student 3:   *"I was thinking about the kinetic energies. Since the velocity is squared to get the kinetic energy but mass isn't, the cart with the bigger speed must have more kinetic energy."*

In the space below, write down whether you agree or disagree with the statements made by each student. Explain.

B.   Which cart takes longer to travel between the two marks? Explain your reasoning.

C.   Using your knowledge of Newton's second law and the definition of acceleration, derive an equation for each cart relating the net force on the cart to the change in velocity of the cart ($\Delta \vec{v}_A$ or $\Delta \vec{v}_B$) and the time interval (Δt_A or Δt_B) that the cart spends between the two marks.

1.   Is the quantity $m_A|\Delta \vec{v}_A|$ *greater than, less than,* or *equal to* $m_B|\Delta \vec{v}_B|$? Explain how you can tell.

For a constant net force, the quantity $\vec{F}_{net} \Delta t$ is called the *impulse* imparted on the object.

2. Is the magnitude of the impulse imparted to cart A *greater than, less than,* or *equal to* the magnitude of the impulse imparted to cart B? Explain your reasoning.

3. Write an equation showing how the impulse imparted to cart A is related to the *change in momentum vector* of cart A $(\Delta \vec{p}_A)$, where momentum, denoted by $\vec{p}$, is the product of the mass and velocity of the object.

This relationship is known as the *impulse-momentum theorem.*

4. Is the magnitude of the final momentum of cart A (p_{Af}) *greater than, less than,* or *equal to* the magnitude of the final momentum of cart B (p_{Bf})? Explain.

D. How does the net work done on cart A $(W_{net, A})$ compare to the net work done on cart B $(W_{net, B})$? Explain.

Compare the kinetic energy of cart A to the kinetic energy of cart B after they have passed the second mark.

E. Refer again to the discussion among the three students in part A. Do you agree with your original answer?

If you disagree with any of the students, identify what is incorrect with their statements.

⇨ Discuss your answers with a tutorial instructor before continuing.

©Prentice Hall
Preliminary Edition, 1998

II. Application of the work-energy and impulse-momentum theorems

Obtain a wedge, a ball, a
cardboard ramp, and
enlargements of the two figures
below.

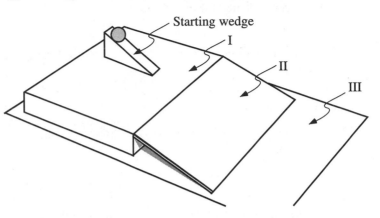

(*Note:* It is important that each
time the ball is rolled it has the
same speed on the level
region I. Place a mark halfway
up the wedge and release the
ball from the mark each time.)
Ignore the rotation of the ball.

A. Release the ball so that it
    rolls straight toward the ramp
    (motion 1).

    Observe the motion of the ball.

    Sketch the remainder of the
    trajectory of the ball on the
    enlargement for motion 1.

    Draw arrows in region II of the
    enlargement that represent the
    direction of the acceleration of the
    ball and the net force on the ball
    while it is on the ramp.

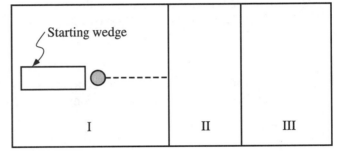

Top view, motion 1

B. Release the ball at an angle to the
    ramp as shown (motion 2).

    Observe the motion of the ball.

    Sketch the remainder of the
    trajectory of the ball on the
    enlargement for motion 2.

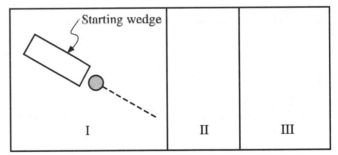

Top view, motion 2

Draw arrows in region II of the enlargement that represent the direction of the acceleration of
the ball and the net force on the ball while it is on the ramp.

How does the direction of the net force on the ball in motion 2 compare to the direction
of the net force on the ball in motion 1? Explain.

Is the direction of the acceleration vector of the ball in motion 2 consistent with the fact
that the ball speeds up and its trajectory curves? Explain.

C. How does the change in kinetic energy of the ball in motion 1 compare to the change in kinetic energy of the ball in motion 2?

 1. Is your answer consistent with the net work done on the ball in motions 1 and 2? Explain your reasoning.

 2. How does the ball's final speed in motion 1 compare to its final speed in motion 2? Explain.

D. For motion 1, draw vectors in region II of the enlargement that represent the momentum of the ball at the top of the ramp and at the bottom of the ramp (at the top and bottom of region II). Use these vectors to construct the change in momentum vector $\Delta \vec{p}$.

 How is the direction of $\Delta \vec{p}$ related to the direction of the net force on the ball as it rolls down the ramp? Is your answer consistent with the impulse-momentum theorem?

E. For motion 2, draw vectors in region II of the enlargement that represent the initial and the final momentum of the ball. Use these vectors to construct the change in momentum vector $\Delta \vec{p}$.

 1. How is your answer in part C regarding the ball's final speed reflected in the final momentum vector that you drew on the enlargement?

 2. How should the direction of $\Delta \vec{p}$ compare to the direction of the net force on the ball as it rolls down the ramp? If necessary, modify your diagram to be consistent with the impulse-momentum theorem.

F. Consider the change in momentum vectors you constructed for motions 1 and 2.

 1. How do they compare in direction? How do they compare in magnitude?

 2. On the basis of your answer, compare the time that the ball spends on the ramp in motion 1 to the time it spends on the ramp in motion 2. Explain. (*Hint:* Can you use the impulse-momentum theorem to compare the time intervals?)

 Is your answer consistent with the trajectory of the ball in each motion? Explain.

Tutorials in Introductory Physics
McDermott, Shaffer, & P.E.G., U.Wash.

©Prentice Hall
Preliminary Edition, 1998

I.   Coupled blocks

Two blocks are on top of a frictionless, level table and are attached to each other by a massless spring. After the blocks have been pulled apart slightly, so that the spring is stretched, the blocks are released from rest at the same time.

$m_A = 3m_B$ $m_S = 0$ Table is frictionless

The mass of block A is three times the mass of block B.

A. Draw a separate free-body diagram for each block and for the spring immediately after release. Clearly label the forces.

Copy your free-body diagrams here after discussion with your partners.

Free-body diagram for block A	Free-body diagram for spring	Free-body diagram for block B

Identify all the Newton's third law force pairs in your diagrams by placing a small "$\times$" through each member of the pair (*i.e.*, mark each member of the first pair as $\rightarrow\!\!\times\!\!\blacktriangleright$, and each member of the second pair as $\times\!\!\times\!\!\blacktriangleright$, *etc.*).

B. Rank the magnitudes of all the *horizontal* forces on your diagrams. If any of the forces have the same magnitude, state that explicitly. Explain.

Is your ranking consistent with the fact that the spring is massless? Explain.

How does the net force on block A compare to the net force on block B? Discuss both magnitude and direction.

Does this comparison hold true:

• immediately after the blocks are released? Explain.

• for all times after the blocks are released? Explain.

Tutorials in Introductory Physics
McDermott, Shaffer, & P.E.G., U.Wash.

©Prentice Hall
Preliminary Edition, 1998

C. The change in velocity vector for block A is shown for a small time interval Δt after the blocks are released. Draw the change in velocity vector for block B over the same time interval. Show the correct relative magnitude.

Explain how Newton's second law and the definition of acceleration can be used to determine the directions and the relative magnitudes of these two vectors.

How does $m_A \Delta \vec{v}_A$ compare to $m_B \Delta \vec{v}_B$ for this small time interval?

Would this comparison change if we considered:

• another interval of time, equally small, occurring much later?

• a much larger time interval?

Explain.

D. The *momentum* of an object is defined to be the product of the mass of the object and the velocity of the object. Conventionally, the momentum is denoted by the symbol $\vec{p}$.

In the space provided, draw the *change in momentum vector* for each block over the same small time interval as in part C. Show the correct relative magnitudes.

Explain how you determined your answers.

How would $\Delta \vec{p}_A$ compare to $\Delta \vec{p}_B$ if we considered a much larger time interval? Explain.

E. Is there some instant after the blocks are released when the velocity of block A becomes zero? If so, is the velocity of block B at this instant also zero? Explain your reasoning.

➪ Discuss your answers with a tutorial instructor before continuing.

Tutorials in Introductory Physics
McDermott, Shaffer, & P.E.G., U.Wash.

©Prentice Hall
Preliminary Edition, 1998

II.   Systems of multiple bodies

Let system C represent the system consisting of blocks A and B
and the spring.

A. Draw and label a free-body diagram for system C. Also
    show separately the net force on system C.

    Which forces in your free-body diagrams in section I do not
    have corresponding forces on the free-body diagram for
    system C?

Free-body diagram for system C

Net force on system C

    Such forces are sometimes called *internal forces,* to be

distinguished from *external forces.*

B. The *momentum of a system* containing multiple bodies can be defined to be the sum of the
    momentum vectors of the constituent bodies.

    Use this definition to write an expression for the momentum of system C.

    Determine the momentum of system C at each of the following times:

    • immediately after the blocks are released,

    • a short time later, when the blocks have undergone the change in momentum indicated in
      section I, and

    • a much longer time later.

    Explain how you determined your answers.

C. Generalize from your results to answer the following question.

    When the *net* force on a system is zero, how does the momentum vector of that system
    behave as time passes?

    When this condition holds, the momentum of the system is said to be *conserved.*

    Is it correct in this case to say that the momentum of block A is conserved? that the
    momentum of block B is conserved? Explain.

D. Imagine a single object whose mass is equal to the mass of system C and whose momentum is equal at all times to $\vec{p}_C$. Draw an arrow that represents the direction of the velocity of that object. If the velocity is zero, state that explicitly.

Direction of $\vec{v}$

The velocity that you have found is called the velocity of the *center of mass*, $\vec{v}_{cm}$, of system C.

III. Disconnected blocks

The spring formerly connecting blocks A and B is disconnected from block B and the blocks are each given a quick push so that they will collide with the spring between them. As above, system C refers to the combination of both blocks and spring.

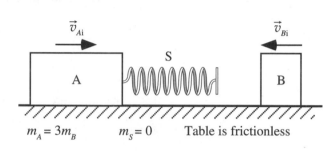

$m_A = 3m_B$ $m_S = 0$ Table is frictionless

A. What are the *external* forces exerted on system C during the collision?

What is the *net force* on system C?

The momentum vectors of each block *before* the collision and of block A *after* the collision are shown below. Complete the table.

	Block A	Block B	System C
Momentum *before* the collision	(arrow right, long)	(arrow left)	
Momentum *after* the collision	(arrow right, short)		
Change in momentum			

How does $\Delta\vec{p}_A$ compare to $\Delta\vec{p}_B$?

How do the final speeds of the blocks compare? Explain.

B. For an instant *before* and *after* the collision, draw an arrow that represents the direction of the velocity of the center of mass of system C.

Direction of $\vec{v}_{cm,i}$ Direction of $\vec{v}_{cm,f}$

As a result of the collision, does the speed of the center of mass of system C *increase, decrease,* or *stay the same?* Explain.

Tutorials in Introductory Physics
McDermott, Shaffer, & P.E.G., U.Wash.

©Prentice Hall
Preliminary Edition, 1998

I. Coupled blocks

Two blocks connected by a massless spring are on top of a frictionless, level table. The blocks are pulled apart slightly so that the spring is stretched, and while they are held apart they are given identical initial velocities in a direction perpendicular to the spring. The blocks are then released at the same time.

The mass of block A is two and a half times the mass of block B.

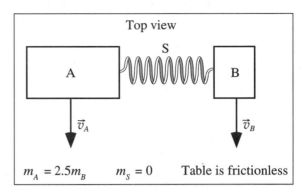

A. Draw separate free-body diagrams for each block and for the spring immediately after release. Indicate separately the *vertical* forces (perpendicular to the table top) and the *horizontal* forces (parallel to the table top). Clearly label all forces.

Copy your free-body diagrams below after discussion.

Free-body diagram for block A	Free-body diagram for spring	Free-body diagram for block B
Vertical forces	Vertical forces	Vertical forces
Horizontal forces	Horizontal forces	Horizontal forces

Identify all the Newton's third law force pairs in your diagrams by placing a small "✕" through each member of the pair (*i.e.,* mark each member of the first pair as →✕▶, and each member of the second pair as ✕✕▶, *etc.*).

B. Rank the magnitudes of all the *horizontal* forces on your diagrams. If any of the horizontal forces have the same magnitude, state that explicitly. Explain how Newton's second and third laws can be used to arrive at this ranking.

How does the net force on block A compare to the net force on block B? Discuss both magnitude and direction.

Does this comparison of the net forces hold true for all times following the release of the blocks? Explain your reasoning.

Tutorials in Introductory Physics
McDermott, Shaffer, & P.E.G., U.Wash.

©Prentice Hall
Preliminary Edition, 1998

C. The velocity vectors for blocks A and B are shown for a time immediately before release. Draw the change in velocity vector for each block for a small time interval Δt after release.

$\vec{v}_{Ai}$	$\vec{v}_{Bi}$	$\Delta\vec{v}_A$	$\Delta\vec{v}_B$

Explain how Newton's second law and the definition of acceleration can be used to determine the directions of the change in velocity vectors.

By what factor is the magnitude of $\Delta\vec{v}_B$ greater than the magnitude of $\Delta\vec{v}_A$? Explain.

How does $m_B\Delta\vec{v}_B$ compare to $m_A\Delta\vec{v}_A$ for this small time interval?

Would this comparison change if we considered:

• another interval of time, equally small, occurring much later?

• a much larger time interval?

Explain.

D. Use your knowledge of the velocities and changes in velocities to construct *momentum vectors* and *change in momentum vectors* for the blocks. Also draw a final momentum vector for each block corresponding to the same small time interval as in part C. Show the correct relative magnitudes.

$\vec{p}_{Ai}$	$\vec{p}_{Bi}$	$\Delta\vec{p}_A$	$\Delta\vec{p}_B$	$\vec{p}_{Af}$	$\vec{p}_{Bf}$

Explain how you determined these vectors.

How would $\Delta\vec{p}_A$ compare to $\Delta\vec{p}_B$ if we considered a much larger time interval? Explain.

Tutorials in Introductory Physics
McDermott, Shaffer, & P.E.G., U.Wash.

©Prentice Hall
Preliminary Edition, 1998

Is there some instant after release when the velocity of block A becomes the same as it was at the time of release? If so, how does the velocity of block A compare to the velocity of block B at this instant? Explain.

II. Coupled blocks treated as a system

Let system C denote the combined system of blocks A and B and the spring S.

A. Draw and label a free-body diagram for system C at a time following the release of the blocks. Indicate separately the *vertical* forces (perpendicular to the table top) and the *horizontal* forces (parallel to the table top). Show separately the net force on system C.

Which forces in your free-body diagrams in section I are *internal forces* for system C?

Free-body diagram for system C	Net force on
Vertical forces	system C
Horizontal forces	

B. Write an equation for the momentum of system C in terms of the momenta of its constituent bodies.

Compare the momentum of system C *immediately* after the blocks are released to its momentum at the following times:

• a short time later, when the blocks have undergone the change in momentum indicated in section I, and

• a much longer time later.

Explain how you determined your answers.

C. Generalize from your results to answer the following question: Under what condition will the momentum of a system be conserved?

D. Describe the motion of the *center of mass* of system C. Explain.

III.   Colliding blocks

The spring formerly connecting blocks A and B is disconnected from block B. The blocks are given initial velocities in the directions shown so that they will collide with the spring between them. As in section II, system C refers to the combination of both blocks and spring.

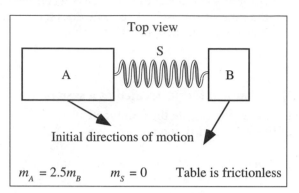

A. What are the *external* forces exerted on system C during the collision?

What is the *net force* on system C?

B. The momentum vectors of each block before the collision and of block B after the collision are shown. Complete the table to show the momentum vectors of system C before and after the collision and of block A after the collision.

	Block A	Block B	System C
Momentum *before* the collision			
Momentum *after* the collision			
Change in momentum			

How do the final speeds of the blocks compare? Explain.

C. Draw arrows that represent the direction of the velocity of the center of mass of system C *before* and *after* the collision.

Direction of $\vec{v}_{cm,\,i}$	Direction of $\vec{v}_{cm,\,f}$

As a result of the collision, does the speed of the center of mass of system C *increase, decrease,* or *stay the same?* Explain.

©Prentice Hall
Preliminary Edition, 1998

Rotation

Tutorials in Introductory Physics
McDermott, Shaffer, & P.E.G., U.Wash.

©Prentice Hall
Preliminary Edition, 1998

I.   Constant angular velocity

A wheel is spinning *counter-clockwise* at a constant rate about a fixed axis. The diagram at right represents a snapshot of the wheel at one instant in time.

A. Draw arrows on the diagram to represent the direction of the velocity for each of the points *A, B,* and *C* at the instant shown. Explain your reasoning.

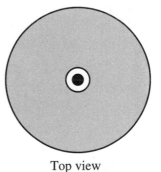

Top view
Wheel spins *counter-clockwise*

Is the time taken by points *B* and *C* to move through one complete circle *greater than, less than,* or *the same as* the time taken by point *A?*

On the basis of your answer above, determine how the speeds of points *A, B,* and *C* compare. Explain.

B. Mark the position of each of the labeled points at a later time when the wheel has completed one half of a turn. Sketch a velocity vector at each point.

For each labeled point, how does the velocity compare to the velocity at the earlier time in part A? Discuss both magnitude and direction.

Top view
Wheel spins *counter-clockwise*

Is there one single *linear velocity vector* that applies to every point on the wheel at all times? Explain.

Tutorials in Introductory Physics
McDermott, Shaffer, & P.E.G., U.Wash.

©Prentice Hall
Preliminary Edition, 1998

C. Suppose the wheel makes one complete revolution in 2 seconds.

    1. For each of the following points, find the change in angle ($\Delta\theta$) of the position vector during one second. (*i.e.,* Find the angle between the initial and final position vectors.)

       • point *A*

       • point *B*

       • point *C*

    2. Find the rate of change in the angle for any point on the wheel.

The rate you calculated above is called the *angular speed* of the wheel, or equivalently, the magnitude of the *angular velocity* of the wheel. The angular velocity is defined to be a vector that points along the axis of rotation and is conventionally denoted by the symbol $\vec{\omega}$ (the Greek letter *omega*). To determine the direction of the angular velocity vector, we imagine an observer on the axis of rotation who is looking toward the object. If the observer sees the object rotating counter-clockwise, the angular velocity vector is directed toward the observer; if the observer sees it rotating clockwise, the angular velocity vector is directed away from the observer.

D. Would two observers on either side of a rotating object agree on the *direction* of the angular velocity vector? Explain.

    Would two observers who use different points on an object to determine the angular velocity agree on the *magnitude* of the angular velocity vector? Explain.

E. The diagrams at right show top and side views of the spinning wheel in part A.

    On each diagram, draw a vector to represent the angular velocity of the wheel. (Use the convention that ⊙ indicates a vector pointing *out of* the page and ⊗ indicates a vector pointing *into* the page.)

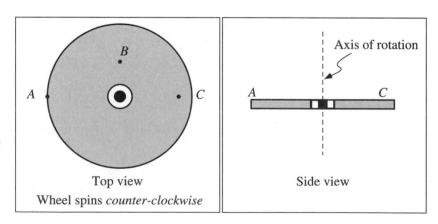

Top view
Wheel spins *counter-clockwise*

Side view

F. In the space at right sketch the position vectors for point *C* at the beginning and at the end of a small time interval Δ*t*.

> Sketch of position vectors at
> t_o and $t_o + \Delta t$

1. Label the change in angle (Δ*θ*) and the distance between the center of the wheel and point *C* (r_C). Sketch the path taken by point *C* during this time interval.

    What is the distance that point *C* travels during Δ*t*? Express your answer in terms of r_C and Δ*θ*.

2. Use your answer above and the definition of linear speed to derive an algebraic expression for the linear speed of point *C* in terms of the angular speed *ω* of the wheel.

    What does your equation imply about the relative linear speeds for points farther and farther out on the wheel? Is this consistent with your answer to part A?

II.   Changing angular velocity

A. Let $\vec{\omega}_o$ represent the initial angular velocity of a wheel. In each case described below, determine the magnitude of the *change in angular velocity* $|\Delta \vec{\omega}|$ in terms of $|\vec{\omega}_o|$.

1. The wheel is made to spin faster, so that eventually, a fixed point on the wheel is going around twice as many times each second. (The axis of rotation is fixed.)

2. The wheel is made to spin at the same rate but in the opposite direction.

B. Suppose the wheel slows down uniformly, so that $|\vec{\omega}|$ decreases by 8π rad/s every 4 s. (The wheel continues spinning in the same direction and has the same orientation.)

    Specify the angular acceleration $\vec{\alpha}$ of the wheel by giving its magnitude and, relative to $\vec{\omega}$, its direction.

    In linear kinematics we find the acceleration vector by first constructing a *change in velocity vector* $\Delta \vec{v}$ and then dividing that by Δ*t*. Describe the analogous steps that you used above to find the angular acceleration $\vec{\alpha}$.

⇨ Discuss your answers above with a tutorial instructor before continuing.

III. Angular acceleration and torque

The rigid bar shown at right is free to rotate about a fixed pivot through its center. The axis of rotation of the bar is perpendicular to the plane of the paper.

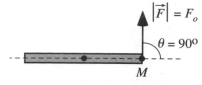

A. A force of magnitude F_o is applied to point M as shown. The force is *always* at a right angle to the bar.

For each of the following cases, determine whether the angular acceleration would be in a *clockwise* or *counter-clockwise* sense.

• The bar was initially at rest. *(Hint: Consider $\Delta\vec{\omega}$.)*

• The bar was spinning at a constant rate before the force was applied.

Does your answer for the angular acceleration depend on whether the bar was originally spinning clockwise or counter-clockwise? Explain.

The application point and direction of a force can affect the rotational motion of the object to which the force is applied. The tendency of a particular force to cause an angular acceleration of an object is quantified as the *torque* produced by the force. The torque $\vec{\tau}$ is defined to be the vector cross product $\vec{r} \times \vec{F}$, where $\vec{r}$ is the vector from the axis of rotation to the point where the force is applied. The magnitude of the torque is simply $|\vec{\tau}| = |\vec{r}|\,|\vec{F}|\sin\theta$, where θ is the angle between $\vec{r}$ and $\vec{F}$.

B. Compare the magnitude of the *net torque* about the pivot in part A to that in each case below.

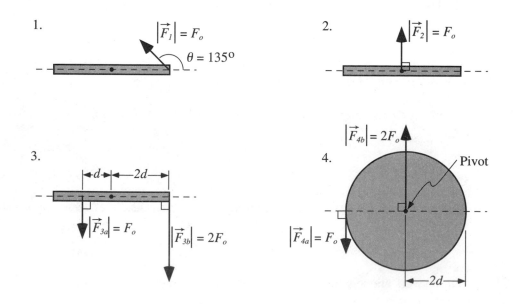

Tutorials in Introductory Physics
McDermott, Shaffer, & P.E.G., U.Wash.

©Prentice Hall
Preliminary Edition, 1998

I.   Extended free-body diagrams

A. A ruler is placed on a pivot and held at an angle as shown at right. The pivot passes through the center of the ruler.

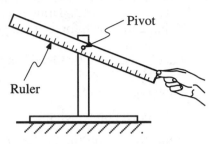

Predict the motion of the ruler after it is released from rest. Explain your reasoning.

Check your prediction by observing the demonstration.

1. Is the angular acceleration of the ruler in a *clockwise sense,* in a *counter-clockwise sense,* or *zero?* Explain how you can tell.

What does your answer imply about the *net torque* on the ruler about the pivot? Explain.

2. What is the direction of the acceleration of the center of mass of the ruler? If $\vec{a}_{cm} = 0$, state that explicitly. Explain how you can tell.

What does your answer imply about the *net force* acting on the ruler? Explain.

Tutorials in Introductory Physics
McDermott, Shaffer, & P.E.G., U.Wash.

©Prentice Hall
Preliminary Edition, 1998

B. Draw a free-body diagram for the ruler (after it is released from rest). Draw your vectors on the diagram at right. Draw each force at the point at which it is exerted.

Label each force by identifying:

- the type of force,

- the object on which the force is exerted, and

- the object exerting the force.

The diagram you have drawn is called an *extended free-body diagram.*

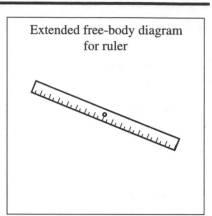

Extended free-body diagram for ruler

Is the point at which you placed the gravitational force in your diagram consistent with your knowledge of the net torque about the pivot? Explain.

C. How would your free-body diagram change if the ruler had twice its original mass (and the same dimensions as before)? Explain.

II.   Net torque and net force

Two identical spools are held the same height above the floor. The thread from spool A is tied to a support, while spool B is not connected to a support. An "×" is marked on the floor directly below each spool.

Both spools are released from rest at the same instant. (Make the approximation that the thread is massless.)

Draw an extended free-body diagram for each spool at an instant after they are released but before they hit the ground.

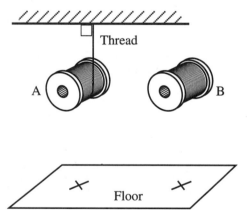

Extended free-body diagram for spool A	Extended free-body diagram for spool B

For each spool, determine the direction of the net torque about the center of the spool. If the net torque is zero, state that explicitly. Explain your reasoning.

Tutorials in Introductory Physics
McDermott, Shaffer, & P.E.G., U.Wash.

©Prentice Hall
Preliminary Edition, 1998

A. *Predict:*

- which spool will reach the floor first. Explain how your answer is consistent with your extended free-body diagrams.

- whether spool A will strike the floor to the left of the "×," to the right of the "×," or fall straight down. Explain how your answer is consistent with your free-body diagrams.

Describe how the net force is related to the individual forces on a free-body diagram when the forces are exerted at different points on the object.

B. Obtain two spools and a ring stand. Use the equipment to check your predictions. (Be sure the thread of spool A is vertical before the spools are released.)

1. How does the *magnitude* of the acceleration of the center of mass $(\vec{a}_{cm})$ of spool A compare to that of spool B? Is this consistent with your free-body diagrams?

2. How does the *direction* of the acceleration of the center of mass $(\vec{a}_{cm})$ of spool A compare to that of spool B? Is this consistent with your free-body diagrams?

C. Below is an *incorrect* statement about spool A.

> *"The string exerts a force that is tangent to the rim of spool A. This force has no component that points toward the center of the spool, so this force does not affect the acceleration of the center of mass."*

Identify the incorrect reasoning in the above statement. Explain.

If necessary, revise your description in part A of how the net force is related to forces exerted at different points on an object.

D. Write down Newton's second law for each spool. Express your answer in terms of the mass of each spool *(m)*, the acceleration of the center of mass of each spool $(\vec{a}_{cm})$ and the net force on each spool $(\vec{F}_{net})$.

Write down the rotational analogue to Newton's second law for each spool. Express your answer in terms of the relevant rotational quantities, that is, in terms of the angular acceleration $(\vec{\alpha})$, the rotational inertia *(I)*, and the torque $(\vec{\tau}_{net})$.

Tutorials in Introductory Physics
McDermott, Shaffer, & P.E.G., U.Wash.

©Prentice Hall
Preliminary Edition, 1998

Electrostatics

Tutorials in Introductory Physics
McDermott, Shaffer, & P.E.G., U.Wash.

I. Electrical interactions

A. Press a piece of sticky tape, about 15-20 cm in length, firmly onto a smooth unpainted surface, for example, a notebook or an unpainted tabletop. (For ease in handling, make "handles" by folding each end of tape to form portions that are not sticky.) Then peel the tape off the table and hang it from a support (*e.g.,* a wooden dowel or the edge of a table).

Describe the behavior of the tape as you bring objects toward it (*e.g.,* a hand, a pen).

B. Make another piece of tape as described above. Bring the second tape toward the first. Describe your observations.

It is important, as you perform the experiment above, that you keep your hands and other objects away from the tapes. Explain why this precaution is necessary.

How does the distance between the tapes affect the interaction between them?

C. Each member of your group should press a tape onto the table and write a *"B"* (for bottom) on it. Then press another tape on top of each *B* tape and label it *"T"* (for top).

Pull each pair of tapes off the table as a unit. After they are off the table, separate the *T* and *B* tapes. Hang one of the *T* tapes and one of the *B* tapes from the support at your table.

Describe the interaction between the following pairs of tape when they are brought near one another.

- two *T* tapes
- two *B* tapes
- a *T* and a *B* tape

Tutorials in Introductory Physics
McDermott, Shaffer, & P.E.G., U.Wash.

©Prentice Hall
Preliminary Edition, 1998

D. Obtain an acrylic rod and a piece of wool. Rub the rod with the wool, and then hold the rod near newly made *T* and *B* tapes on the wooden dowel.

Compare the interactions of the rod with the tapes to the interactions between the tapes in part C. Describe any similarities or differences.

We say that the rod and tapes are *electrically charged* when they interact as you have observed.

E. Base your answers to the following questions on the observations you have made thus far.

1. How many different types of charge do there appear to be? Explain.

2. By convention, the acrylic rod is said to be positively charged when rubbed with wool. How do two objects that are positively charged interact? Explain how you can tell.

3. Which tape, *T* or *B,* has a positive charge? Explain.

Please remove all tape from the tabletop before continuing.

II. Superposition

A. Obtain a small pith ball attached to an insulating string.

Touch the ball to a charged rod and observe the behavior of the ball after it touches the rod.

Is the ball charged? If so, does the ball have the same charge or opposite charge as the rod? Explain how you can tell.

Tutorials in Introductory Physics
McDermott, Shaffer, & P.E.G., U.Wash.

©Prentice Hall
Preliminary Edition, 1998

B. Hold a charged rod as shown. (The top view shows a plane through the center of the rod.)

Explore the region near the rod with a test ball that has a charge of the same sign as the rod.

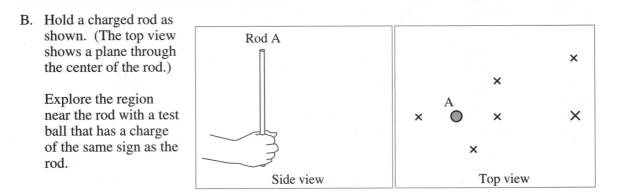

Based on your observations, sketch *vectors* to represent the net electric force on the ball at each of the points marked by an "x."

C. Hold an identically charged rod about 15 cm to the right of the original rod and repeat the experiment in part B.

Sketch *vectors* at the marked points to represent the net electric force exerted on the ball.

1. Discuss with your partners what is meant by the principle of superposition.

2. Are your results consistent with the principle of superposition? Explain.

3. Does rod A exert a force on the test charge when the test charge is *directly* to the right of rod B (*e.g.,* at the point marked by the large "X")? Explain.

Check your answer by observing the test charge at this point while rod A is moved toward rod B.

III. A model for electric charge

A. A small ball with zero net charge is positively charged on one side, and equally negatively charged on the other side. The ball is placed near a positive point charge as shown.

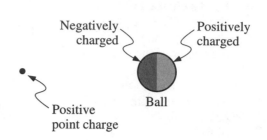

Would the ball be *attracted toward, repelled from,* or *unaffected by* the positive point charge? Explain.

Is your answer consistent with Coulomb's law? Explain.

B. Hang an uncharged metal or metal-covered ball from an insulating string. Then charge a piece of tape as in part A of section I and bring the tape toward the ball.

Describe what you observe.

Through careful observations of physical phenomena, scientists develop models, or mental pictures, to account for observations. These scientific models can also be used to predict physical behavior. From observations of electrical phenomena we can develop a model for electric charge.

C. Use your model for electric charge to account for the electrical attraction between charged tape and an uncharged metal ball. As part of your answer, draw a sketch of the charge distribution on the tape and ball both before and after they are brought near one another.

D. Two students discuss what would happen if, instead of being uncharged, the conducting ball in part B had a small net charge. (Assume the ball and the tape are both positively charged.)

Student 1: *"Since the ball is a conductor, the excess charge will be evenly distributed on the surface. Since the ball and the tape have like charges, they will repel."*

Student 2: *"The tape will still repel the positive charge on the ball and attract the negative charge on the ball. If the excess positive charge is not too much they may still attract each other."*

Student 1: *"That can't be true. If the ball has a net positive charge then there is no negative charge on the ball."*

Do you agree with either student? Explain your reasoning.

E. Two identical metal balls are mounted on stationary insulating stands. Initially, the balls have the same non-zero net charge.

Ball 1   Ball 2

1. Draw a separate free-body diagram for each ball. Identify any third law force pairs (for example, mark each member of the first pair as ⟶✕▶, and each member of the second pair as ✕✕▶, *etc.*).

Label each of the forces to indicate:

- the object exerting the force,
- the object on which the force is exerted,
- the type of force (gravitational, normal, *etc.*), and
- whether the force is a *contact* or a *non-contact* force.

Are there any forces represented on your diagrams that are equal in magnitude? Explain.

Tutorials in Introductory Physics
McDermott, Shaffer, & P.E.G., U.Wash.

©Prentice Hall
Preliminary Edition, 1998

2. Suppose that the charge on ball 2 is decreased so that it is less than that on ball 1. How do the free-body diagrams for the balls in this case compare to the free-body diagrams that you drew above? Explain.

3. Suppose that the net charge on ball 2 is reduced to zero. How do the free-body diagrams for this case compare to the diagrams you drew in parts 1 and 2 above? Explain.

    Is your answer consistent with the observations you made of an uncharged ball when a charged tape was brought near it? (See part B.) If not, modify your answer so that it is consistent.

Tutorials in Introductory Physics
McDermott, Shaffer, & P.E.G., U.Wash.

©Prentice Hall
Preliminary Edition, 1998

I. Area as a vector

A. Hold a small piece of paper (*e.g.*, an index card) flat in front of you. The paper can be thought of as a part of a larger plane surface.

What *single* line could you use to specify the orientation of the plane of the paper (*i.e.*, so that someone else could hold the paper in the same, or in a parallel, plane)?

B. The area of a flat surface can be represented by a single *vector*, called the area vector $\vec{A}$.

What does the direction of the vector represent?

What would you expect the magnitude of the vector to represent?

C. Place a large piece of graph paper flat on the table.

Describe the direction and magnitude of the area vector, $\vec{A}$, for the entire sheet of paper.

Describe the direction and magnitude of the area vector, $d\vec{A}$, for each of the individual squares that make up the sheet.

D. Fold the graph paper twice so that it forms a hollow triangular tube.

Can the entire sheet be represented by a single vector with the characteristics you defined above? If not, what is the minimum number of area vectors required?

E. Form the graph paper into a tube as shown.

Can the orientation of each of the individual squares that make up the sheet of graph paper still be represented by $d\vec{A}$ vectors as in part C above? Explain.

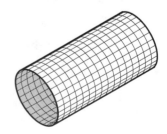

F. What must be true about a surface or a portion of a surface in order to be able to associate a single area vector $\vec{A}$ with that surface?

Tutorials in Introductory Physics
McDermott, Shaffer, & P.E.G., U.Wash.

©Prentice Hall
Preliminary Edition, 1998

II. Electric field

A. In the *Charge* tutorial, you explored the region around a charged rod with a pith ball that had a charge of the same sign as the rod.

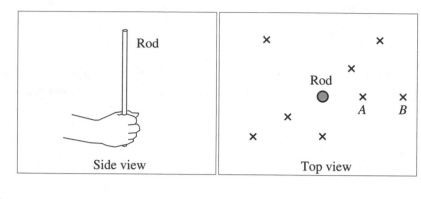

Side view

Top view

Sketch *vectors* at each of the marked points to represent the electric force exerted on the ball at that location.

How does the magnitude of the force exerted on the ball at point *A* compare to the magnitude of the force on the ball at point *B?*

B. Suppose that the charge, q_{test}, on the pith ball were halved.

Would the electric force exerted on the ball at each location change? If so, how? If not, explain why not.

Would the ratio $\vec{F} / q_{test}$ change? If so, how? If not, explain why not.

C. The quantity $\vec{F} / q_{test}$ evaluated at any point is called the *electric field* $\vec{E}$ at that point.

How does the magnitude of the electric field at point *A* compare to the magnitude of the electric field at point *B?* Explain.

D. Sketch *vectors* at each of the marked points to represent the electric field $\vec{E}$ at that location.

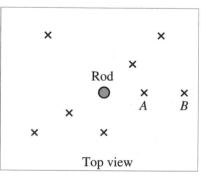

Top view

Would the magnitude or the direction of the electric field at point *A* change if:

- the charge on the *rod* were increased? Explain.

- the magnitude of the test charge were increased? Explain.

- the sign of the test charge were changed? Explain.

©Prentice Hall
Preliminary Edition, 1998

The electric field can be represented in two ways: either with *vectors,* as in your preceding drawing, or alternatively with *field lines*. When drawn in this way, the *direction* of the electric field at any point is tangent to the electric field line through that point, and the lines are directed away from positive charges and toward negative charges. Next we explore how the field line representation can reflect the *magnitude* of the electric field as well.

E. The diagram at right shows a two-dimensional top view of the *electric field lines* representing the electric field for a positively charged rod.

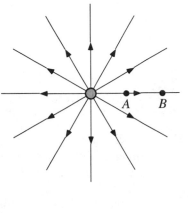

You determined previously that the magnitude of the electric field at point *A* was larger than the field at point *B*. What feature of the *electric field lines* reflects this information about the magnitude of the field?

III. Flux

Ask a tutorial instructor for a block of wood with nails through it. The nails represent uniform electric field lines. (The block of wood does not represent anything but serves to hold the nails in place.)

At right is a two-dimensional representation of the same electric field as viewed from the side.

A. Compare the magnitude of the electric field at points *P* and *Q*. Explain your reasoning.

Suppose you were given another block of wood with nails representing a weaker uniform electric field than the one above. How would the two blocks differ? Explain.

B. Obtain a wire loop. The loop represents the *boundary* of an imaginary flat surface of area *A*. (In order to allow the nails that represent the field to pass through the surface, you have only been given the boundary of the surface.)

Draw a diagram to show the relative orientation of the loop and the electric field so that the number of field lines that pass through the surface of the loop is:

• the maximum possible.

• the minimum possible.

For a given surface, the *electric flux, Φ_E*, is proportional to the number of field lines through the surface. At its maximum, the electric flux is equal to the product of electric field at the surface and the surface area *(EA)*. The electric flux is defined to be positive when the electric field $\vec{E}$ has a component in the same direction as the area vector $\vec{A}$.

C. Sketch vectors $\vec{A}$ and $\vec{E}$ such that the electric flux is:

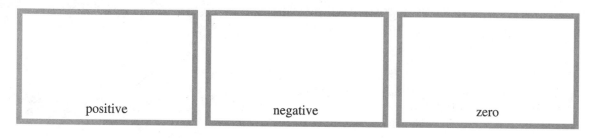

| positive | negative | zero |

D. You will now examine the relationship between the number of field lines through a surface and the angle between $\vec{A}$ and $\vec{E}$.

(You will need a protractor to measure angles.)

1. Place the loop over the nails so that the number of field lines through it is a maximum. Determine the angle between $\vec{A}$ and $\vec{E}$. Record both that angle and the number of field lines that pass through the loop.

2. Rotate the loop until there is one less row of nails passing through it. Determine the angle between $\vec{A}$ and $\vec{E}$ and record your measurement. Continue in this way until $\theta = 180°$.

3. On graph paper, plot a graph of *n* versus θ. (Let the number of field lines through the surface be a negative number for angles between 90° and 180°.)

n (# of field lines through surface)	θ (angle between $\vec{A}$ and $\vec{E}$)

E. When $\vec{E}$ and $\vec{A}$ are parallel, we called the quantity *EA* the electric flux through the surface. For the parallel case, we found that *EA* is proportional to the number of field lines through the surface.

By what trigonometric function of θ must you multiply *EA* so that the product is proportional to the number of field lines through the area for any orientation of the surface?

Rewrite the quantity described above as a product of just the vectors $\vec{E}$ and $\vec{A}$.

I. Electric flux through closed surfaces

In the tutorial *Electric field and flux* and in the homework, we found that the electric flux through a set of imaginary surfaces, $d\vec{A}_i$, each with a uniform electric field, $\vec{E}_i$, can be written as:

$$\Phi_{net} = \vec{E}_1 \bullet d\vec{A}_1 + \vec{E}_2 \bullet d\vec{A}_2 + \vec{E}_3 \bullet d\vec{A}_3 + \ldots$$

The area vectors at each point on a *closed surface* (*i.e.*, a surface that surrounds a region so that the only way out of the region is through the surface) are chosen by convention to point *out of* the enclosed region. A closed imaginary surface is called a *Gaussian surface.*

A Gaussian cylinder has a radius a and a length l. The end caps are labeled A and C and the side surface is labeled B.

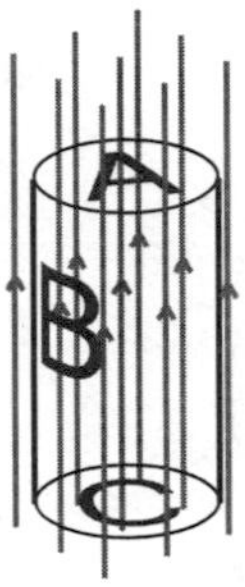

A. The Gaussian cylinder is in a uniform electric field of magnitude E_o that is aligned with the cylinder axis.

- Find the sign and magnitude of the flux through:
  Surface A: Surface B: Surface C:
- Is the net flux through the Gaussian surface *positive, negative,* or *zero?*

B. The Gaussian cylinder encloses a negative charge. (The field from part A is removed.)

- Find the sign of the flux through:
  Surface A: Surface B: Surface C:
- Is the net flux through the Gaussian surface *positive, negative,* or *zero?*

C. The Gaussian cylinder encloses opposite charges of equal magnitude. (The charges are on the axis of the cylinder and equidistant from the center.)

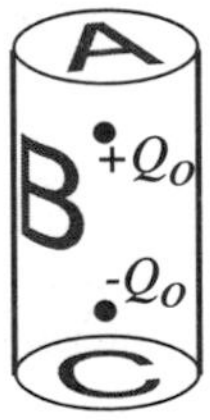

- Find the sign of the flux through:
  Surface A: Surface B: Surface C:
- Is the net flux through the Gaussian surface *positive, negative,* or *zero?*

D. A positive charge is located above the Gaussian cylinder.

- Find the sign of the flux through:
  Surface A: Surface B: Surface C:
- Can you tell whether the net flux through the Gaussian surface is *positive, negative,* or *zero?* Explain.

➪ Check your results with a tutorial instructor before continuing.

II.   Gauss' law

Gauss' law ($\Phi_E = q_{enclosed}/\varepsilon_o$) states that the electric flux through a Gaussian surface is directly proportional to the net charge enclosed by the surface.

A.   Are your answers to parts A-C of section I consistent with Gauss' Law? Explain.

B.   In part D of section I, you tried to determine the sign of the flux through the Gaussian cylinder shown.

   1.   If you have not done so already, use Gauss' law to determine whether the net flux through the Gaussian surface is *positive, negative,* or *zero.* Explain.

   2.   If $\Phi_A = -10$ Nm²/C and $\Phi_C = 2$ Nm²/C, what is Φ_B?

C.   Find the net flux through each of the Gaussian surfaces below.

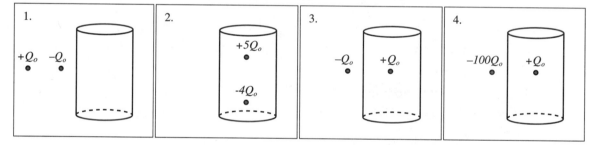

D.   The three spherical Gaussian surfaces at right each enclose a charge $+Q_o$. In case C there is another charge of $-6Q_o$ outside the surface.

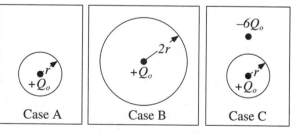

Consider the following conversation:

Student 1:   *"Since each Gaussian surface encloses the same charge, the net flux through each must be the same."*

Student 2:   *"Gauss' law doesn't apply here. The electric field at the Gaussian surface in case B is weaker than in case A, because the surface is farther from the charge. Since the flux is proportional to the electric field strength, the flux must also be smaller in case B."*

Student 3:   *"I was comparing A and C. In C the charge outside changes the field over the whole surface. The areas are the same, so the flux must be different."*

Do you agree with any of the students? Explain.

Tutorials in Introductory Physics
McDermott, Shaffer, & P.E.G., U.Wash.

©Prentice Hall
Preliminary Edition, 1998

III. Application of Gauss' law

A. A large sheet has charge density $+\sigma_o$. A cylindrical
Gaussian surface encloses a portion of the sheet and
extends a distance L_o on either side of the sheet. A_1, A_2,
and A_3 are the areas of the ends and curved side,
respectively. Only a small portion of the sheet is shown.

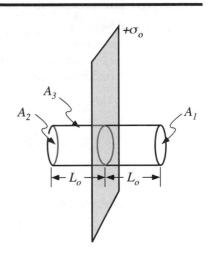

1. On the diagram at right indicate the location of the
    charge enclosed by the Gaussian cylinder.

    In terms of σ_o and other relevant quantities, what is the
    net charge enclosed by the Gaussian cylinder?

2. Sketch the electric field lines on both sides of the sheet.

    Does the Gaussian cylinder affect the field lines or the
    charge distribution? Explain.

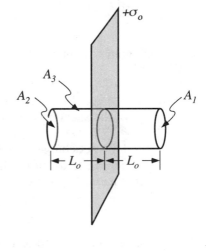

3. Let E_L and E_R represent the magnitude of the electric
    field on the left and right ends of the Gaussian surface.

    How do the magnitudes of E_L and E_R compare?
    Explain.

    How do the magnitudes of the areas of the ends of the Gaussian surface compare?

4. Through which of the surfaces (A_1, A_2, A_3) is there a net flux? Explain using a sketch
    showing the relative orientation of the electric field vector and the area vectors.

    Write an expression for the net electric flux Φ_{net} through the cylinder in terms of the three
    areas $(A_1, A_2,$ and $A_3)$, E_L, and E_R.

    Use the relationships between the electric fields E_L and E_R and between the areas A_1 and
    A_2 to simplify your equation for the net flux.

5. Gauss' law ($\Phi_E = q_{enclosed}/\varepsilon_o$) relates the net electric flux through a Gaussian surface (which you found in part 4) to the net charge enclosed (which you found in part 1). Use this relationship to find the direction and magnitude of the electric field at the right end of the cylinder in terms of σ_o.

What is the electric field at the left end of the cylinder?

Does the electric field near a large sheet of charge depend on the distance from the sheet? Use your results above to justify your answer.

Is your answer consistent with the electric field lines you sketched in part 2? Explain.

⇨ Check your results with a tutorial instructor before you continue.

B. The Gaussian cylinder below encloses a portion of two identical large sheets. The charge density of the sheet on the left is $+\sigma_o$; the charge density of the sheet on the right is $+2\sigma_o$.

1. Find the net charge enclosed by the Gaussian cylinder in terms of σ_o and any relevant dimensions.

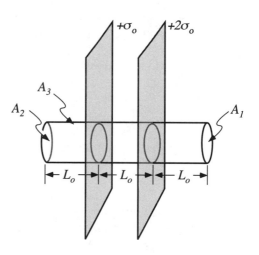

2. Let E_L and E_R be the magnitudes of the electric fields at the left and right end caps of the Gaussian cylinder respectively.

Is E_L *greater than, less than,* or *equal to* E_R? Explain.

3. Find the net flux through the Gaussian cylinder in terms of E_L, E_R, and any relevant dimensions.

4. Use Gauss' law to find the electric field a distance L_o to the right of the rightmost sheet.

Are your results consistent with the results you would obtain using superposition? Explain.

Tutorials in Introductory Physics
McDermott, Shaffer, & P.E.G., U.Wash.

©Prentice Hall
Preliminary Edition, 1998

I. Review of work

A. Suppose an object moves under the influence of a force. Sketch arrows showing the relative directions of the force and displacement when the work done by the force is:

positive	negative	zero

B. An object travels from point A to point B while two constant forces, $\vec{F_1}$ and $\vec{F_2}$, of equal magnitude are exerted on it.

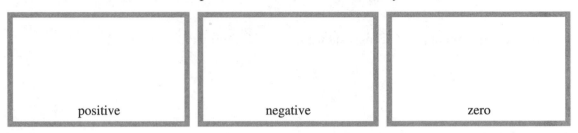

1. Is the total work done on the object by $\vec{F_1}$ *positive, negative,* or *zero?*

2. Is the total work done on the object by $\vec{F_2}$ *positive, negative,* or *zero?*

3. Is the net work done on the object *positive, negative,* or *zero?* Explain.

4. Is the magnitude of the velocity of the object at point B *greater than, less than,* or *equal to* the velocity of the object at point A? Explain how you can tell.

C. An object travels from point A to point B while two constant forces, $\vec{F_3}$ and $\vec{F_4}$, of *unequal* magnitude are exerted on it.

1. Is the total work done on the object by $\vec{F_3}$ *positive, negative,* or *zero?*

2. Is the total work done on the object by $\vec{F_4}$ *positive, negative,* or *zero?*

3. Is the net work done on the object *positive, negative,* or *zero?* Explain.

4. Is the magnitude of the velocity of the object at point B *greater than, less than,* or *equal to* the velocity of the object at point A? Explain how you can tell.

Tutorials in Introductory Physics
McDermott, Shaffer, & P.E.G., U.Wash.

©Prentice Hall
Preliminary Edition, 1998

D. State the work-energy theorem in your own words. Are your answers in part B consistent
with this theorem? Explain.

Are your answers in part C consistent with the work-energy theorem? Explain.

II.   Work and electric fields

The diagram at right shows a top view of a positively
charged rod. Points *W, X, Y,* and *Z* lie in a plane near the
center of the rod. Points *W* and *Y* are equidistant from the
rod, as are points *X* and *Z*.

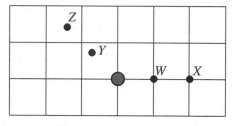

A. Draw electric field vectors at points *W, X, Y,* and *Z*.

B. A particle with charge $+q_o$ travels along a straight line
path from point *W* to point *X*.

Is the work done *by the electric field* on the particle *positive, negative,* or *zero?* Explain using
a sketch that shows the electric force on the particle and the displacement of the particle.

Compare the work done by the electric field when the particle travels from point *W* to point *X*
to that done when the particle travels from point *X* to point *W*.

C. The particle travels from point *X* to point *Z* along the circular arc
shown.

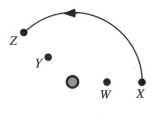

1.   Is the work done *by the electric field* on the particle *positive,*
negative, or *zero?*   Explain. (*Hint:* Sketch the direction of the
force on the particle and the direction of the displacement for
several short intervals during the motion.)

2. Compare the work done by the electric field when the particle travels from point W to point X to that done when the particle travels from point W to point Z along the path shown. Explain.

D. Suppose the particle travels from point W to point Y along the path *WXZY* as shown.

1. Compare the work done by the electric field when the particle travels from point W to point X to that done when the particle travels from point Z to point Y. Explain.

What is the total work done on the particle by the electric field as it moves along the path *WXZY*?

2. Suppose the particle travels from W to Y along the arc shown. Is the work done on the particle by the electric field *positive*, *negative*, or *zero*? Explain using force and displacement vectors.

3. Suppose the particle travels along the straight path *WY*. Is the work done on the particle by the electric field *positive*, *negative*, or *zero*? Explain using force and displacement vectors. (*Hint:* Compare the work done along the first half of the path to the work done along the second half.)

E. Compare the work done as the particle travels from W to Y along the three different paths in part D.

It is often said that the work done by a static electric field is *path independent*. Explain how your results in part D are consistent with this statement.

III. Electric potential difference

A. Suppose the charge of the particle in section II is increased from $+q_o$ to $+1.7q_o$.

1. Is the work done by the electric field as the particle travels from W to X *greater than, less than,* or *equal to* the work done by the electric field on the original particle? Explain.

2. How is the quantity *the work divided by the charge* affected by this change?

The *electric potential difference* ΔV_{WX} between two points W and X is defined to be:

$$\Delta V_{WX} = -\frac{W_{elec}}{q}$$

where W_{elec} is the work done by the field as a charge q travels from W to X.

3. Does this quantity depend on the *magnitude* of the charge of the particle that is used to measure it? Explain.

4. Does this quantity depend on the *sign* of the charge of the particle that is used to measure it? Explain.

Tutorials in Introductory Physics
McDermott, Shaffer, & P.E.G., U.Wash.

©Prentice Hall
Preliminary Edition, 1998

B. Shown at right are four points near a positively
    charged rod. A particle with charge $|q_o| = 2 \times 10^{-6}$ C
    and mass $m_o = 3 \times 10^{-8}$ kg is released from rest at
    point *W*. The speed of the particle is measured to be
    40 m/s as it passes point *X*.

    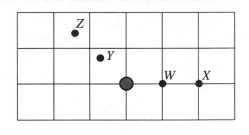

    1. Is q_o *positive* or *negative?* Explain.

    2. Find the change in kinetic energy of the particle as it travels from point *W* to point *X*.

    3. Find the work done on the particle by the electric field between points *W* and *X*.
        (*Hint:* See part D of section I.)

    4. Find the electric potential difference between points *W* and *X*.

C. Consider how the following changes would affect the motion of the particle.

    1. If the same particle were released from point *Y*, would its speed as it passes point *Z* be
        greater than, less than, or *equal to* 40 m/s? Explain.

    2. How, if at all, would your answer to question 4 of part B change if the particle had nine
        times the charge?

    3. How, if at all, would the speed of the particle as it passes point *X* change if the particle
        had nine times the charge?

D. With what speed would you launch another particle with charge $|q_o| = 2 \times 10^{-6}$ C and mass $m_o = 3 \times 10^{-8}$ kg at point Z toward the rod so that it will turn around at point Y? Explain.

E. Suppose you had a particle with the same mass as the particle in part D but with nine times the charge. With what speed would you launch the new particle at point Z toward the rod so that the particle will turn around at point Y? Explain.

Tutorials in Introductory Physics
McDermott, Shaffer, & P.E.G., U.Wash.

©Prentice Hall
Preliminary Edition, 1998

I. The electric field near conducting plates

A. A small portion near the center of a large thin conducting *plate* is shown magnified at right. The portion shown has a net charge Q_1 and each side has an area A_1.

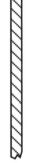

Side view of thin
charged plate

Write an expression for the charge density on each side of the conducting plate.

B. Use the principle of superposition to find the electric field inside the conductor (if you have not done so already).

Is your answer consistent with your knowledge of the electric field inside a conductor? Explain.

C. Use the principle of superposition to find the electric field on each side of the plate.

Does the charge on the *right* surface contribute to the electric field to the *left* of the plate (even though metal separates the two regions)? Explain.

Tutorials in Introductory Physics
McDermott, Shaffer, & P.E.G., U.Wash.

©Prentice Hall
Preliminary Edition, 1998

D. Suppose we had a portion of a large *sheet* of charge also with net charge Q_I and area A_I.

How does the charge density σ' on this sheet compare to the charge density on each side of the plate above? Explain.

How does the electric field on one side of the *sheet of charge* compare to the electric field on the same side of the *charged plate?*

E. A second plate of equal and opposite charge is now held near the first. The plates are large and close enough together that fringing effects near the edges can be ignored.

The diagrams below show various distributions of charge on the two plates. Decide which arrangement is physically correct. Explain.

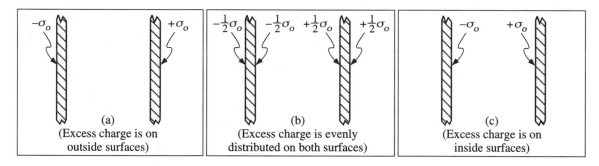

Tutorials in Introductory Physics
McDermott, Shaffer, & P.E.G., U.Wash.

©Prentice Hall
Preliminary Edition, 1998

II.　Parallel plates and capacitance

Two very large thin conducting plates are a distance D apart.
The surface area of the face of each plate is A_o. A side view of a
small portion near the center of the plates is shown.

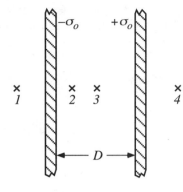

A. The inner surface of one plate has a uniform charge density
   of $+\sigma_o$; the other, $-\sigma_o$. The charge density on the outer
   surface of each plate is zero.

   1. Draw vectors at each labeled point to represent the
      electric field at that point due to each charged plate.

   2. Write expressions for the following quantities in terms of
      the given variables:

      • the electric field at points *1, 2, 3,* and *4.*

      • the potential difference between the plates.

   3. The right plate is moved to the left. Both plates are kept
      insulated. Describe how each of the following quantities
      will change (if at all). Explain.

      • the charge density on each plate

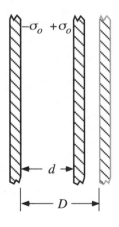

      • the electric field both outside and between the plates

      • the potential difference between the plates

---

4. Write expressions for the following quantities in terms of σ_o and d (the new distance between the plates).

    *   the magnitude of the electric field between the plates

    *   the potential difference between the plates

5. Find $\dfrac{Q}{\Delta V}$ (the ratio of the net charge on one plate to the potential difference between the plates).

    How, if at all, would this ratio change if the charge densities on the plates were $+2\sigma_o$ and $-2\sigma_o$?

⇨ Check your results with a tutorial instructor before you continue.

B. Suppose the plates are discharged, then held a distance D apart and connected to a battery. (Ignore the fringing fields near the plate edges.)

    1. Write expressions for the following quantities. Explain your reasoning in each case.

        *   the potential difference ΔV between the plates

        *   the electric field at points *1, 2, 3,* and *4*

        *   the charge density on each plate

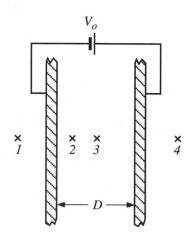

Tutorials in Introductory Physics
McDermott, Shaffer, & P.E.G., U.Wash.

©Prentice Hall
Preliminary Edition, 1998

2. The right plate is moved to the left. Describe how each of the following quantities change (if at all). Explain.

    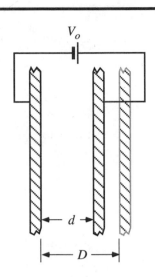

    • the potential difference ΔV between the plates

    • the electric field both outside and between the plates

    • the charge density on each plate

3. Write expressions for the following quantities in terms of V_o and d (the new distance between the plates).

    • the magnitude of the electric field between the plates

    • the charge density on each plate

4. Find $\dfrac{Q}{\Delta V}$ (the ratio of the net charge on one plate to the potential difference between the plates).

    How, if at all, would this ratio change if the voltage of the battery was $2V_o$?

⇨ Check your results with a tutorial instructor before you continue.

Tutorials in Introductory Physics   ©Prentice Hall
McDermott, Shaffer, & P.E.G., U.Wash.   Preliminary Edition, 1998

C. Compare the ratio $\dfrac{Q}{\Delta V}$ that you calculated for two insulated plates (part A) to the same ratio for two plates connected to a battery (part B).

    1. Does the ratio $\dfrac{Q}{\Delta V}$ depend on whether or not the plates are connected to a battery?

    2. Does the ratio $\dfrac{Q}{\Delta V}$ depend on the distance between the plates?

For two plates with equal and opposite charge, the ratio of the charge on one plate to the potential difference between the plates is defined as the *capacitance* ($C = \dfrac{Q}{\Delta V}$).

D. Consider your results from parts A and B.

    1. When the plates are insulated, which quantity or quantities remain fixed and which quantity or quantities change when the distance between the plates changes?

    2. When the plates are connected to a battery, which quantity or quantities remain fixed and which quantity or quantities change when the distance between the plates changes?

Tutorials in Introductory Physics
McDermott, Shaffer, & P.E.G., U.Wash.

©Prentice Hall
Preliminary Edition, 1998

Electric Circuits

Tutorials in Introductory Physics
McDermott, Shaffer, & P.E.G., U.Wash.

In this tutorial, we construct a model for electric current that we can use to predict and explain the behavior of simple electric circuits.

I. Complete circuits

A. Obtain a battery, a light bulb, and a single piece of wire. Connect these in a variety of ways. Sketch each arrangement below.

Arrangements that *do* light the bulb	Arrangements that *do not* light the bulb

You should have found at least four different arrangements that light the bulb. How are these arrangements similar? How do they differ from arrangements in which the bulb does not light?

State the requirements that must be met in order for the bulb to light.

B. A student has briefly connected a wire across the terminals of a battery until the wire feels warm. The student finds that the wire seems to be equally warm at points *1, 2,* and *3.*

Based on this observation, what might you conclude is happening in the wire at one place compared to another?

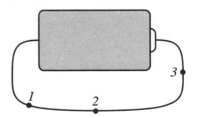

Tutorials in Introductory Physics
McDermott, Shaffer, & P.E.G., U.Wash.

©Prentice Hall
Preliminary Edition, 1998

C. Light a bulb usin g a battery and a single wire. Observe and record the behavior (*i.e.,* brightness) of the bulb when objects made out of various materials are inserted into the circuit. (Try materials such as paper, coins, pencil lead, eraser, your finger, *etc.*)

What is similar about most of the objects that let the bulb light?

D. Carefully examine a bulb. Two wires extend from the filament of the bulb into the base. You probably cannot see into the base, however, you should be able to make a good guess as to where the wires are attached. Describe where the wires attach. Explain based on your observations in parts A–C.

On the basis of the observations that we have made, we will make the following assumptions:

1. A flow exists in a complete circuit from one terminal of the battery, through the rest of the circuit, back to the other terminal of the battery, through the battery and back around the circuit. We will call this flow *electric current.*

2. For identical bulbs, bulb brightness can be used as an indicator of the amount of current through that bulb: the brighter the bulb, the greater the current.

Starting with these assumptions, we will develop a model that we can use to account for the behavior of simple circuits. The construction of a scientific model is a step-by-step process in which we specify only the minimum number of attributes that are needed to account for the phenomena under consideration.

II. Bulbs in series

Set up a two-bulb circuit with identical bulbs connected one after the other as shown. Bulbs connected in this way are said to be connected in *series*.

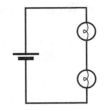

A. Compare the brightness of the two bulbs with each other. (Pay attention only to large differences in brightness. You may notice minor differences if two "identical" bulbs are, in fact, not quite identical.)

Use the assumptions we have made in developing our model for electric current to answer the following questions:

1. Is current "used up" in the first bulb, or is the current the same through both bulbs?

2. Do you think that switching the order of the bulbs might make a difference? Check your answer.

3. On the basis of your observations *alone,* can you tell the direction of the flow through the circuit?

B. Compare the brightness of each of the bulbs in the two-bulb series circuit with that of a bulb in a single-bulb circuit.

Use the assumptions we have made in developing our model for electric current to answer the following questions:

1. How does the current through a bulb in a single-bulb circuit compare with the current through the same bulb when it is connected in series with a second bulb? Explain.

2. What does your answer to question 1 imply about how the current through the *battery* in a single-bulb circuit compares to the current through the *battery* in a two-bulb series circuit? Explain.

C. We may think of a bulb as presenting an obstacle, or *resistance,* to the current in the circuit.

   1. Thinking of the bulb in this way, would adding more bulbs in series cause the total obstacle to the flow, or *total resistance,* to increase, decrease, or stay the same as before?

   2. Formulate a rule for predicting how the current through the battery would change (*i.e.,* whether it would *increase, decrease,* or *remain the same*) if the number of bulbs connected in series were increased or decreased.

III. Bulbs in parallel

Set up a two-bulb circuit with identical bulbs so that their terminals are connected together as shown. Bulbs connected together in this way are said to be connected in *parallel.*

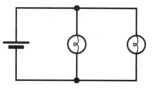

A. Compare the brightness of the bulbs in this circuit.

   1. What can you conclude from your observation about the amount of current through each bulb?

   2. Describe the current in the entire circuit. Base your answer on your observations. In particular, how does the current through the battery seem to divide and recombine at the junctions of the two parallel branches?

Tutorials in Introductory Physics
McDermott, Shaffer, & P.E.G., U.Wash.

©Prentice Hall
Preliminary Edition, 1998

B. Is the brightness of each bulb in the two-bulb parallel circuit *greater than, less than,* or *equal to* that of a bulb in a single-bulb circuit?

How does the amount of current through a *battery* connected to a single bulb compare to the current through a *battery* connected to a two-bulb parallel circuit? Explain based on your observations.

C. Formulate a rule for predicting how the current through the battery would change *(i.e., whether it would increase, decrease, or remain the same)* if the number of bulbs connected in parallel were increased or decreased. Base your answer on your observation of the behavior of the two-bulb parallel circuit and the model for current.

What can you infer about the total resistance of a circuit as the number of parallel branches is increased or decreased?

D. Does the amount of current through a battery seem to depend on the number of bulbs in the circuit and how they are connected?

E. Unscrew one of the bulbs in the two-bulb parallel circuit. Does this change significantly affect the current through the branch that contains the other bulb?

A characteristic of an *ideal* battery is that the branches connected directly across it are independent of one another.

IV. Limitations: The need to extend the model

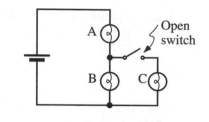

A. The circuit at right contains three identical bulbs and an ideal battery. Assume that the resistance of the switch, when closed, is negligible. Use the model we have developed to:

- predict the relative brightness of the bulbs in the circuit with the switch closed. Explain.

- predict how the brightness of bulb A changes when the switch is opened. Explain.

B. Show that a simple application of the model for current that we have developed thus far is inadequate for determining how the brightness of bulb B changes when the switch is opened.

I.   Current and resistance

A. The circuits at right contain identical batteries, bulbs and unknown identical elements labeled X.

How do the bulbs compare in brightness? Explain.

In each circuit, how does the current through the bulb compare to the current through the element X? Explain.

B. The circuits at right contain identical batteries and bulbs. The boxes labeled X and Y represent different unknown elements. (Assume there are no batteries in either box.)

It is observed that the bulb on the left is brighter than the bulb on the right.

1. Based on this observation, how does the resistance of element X compare to that of element Y? Explain.

2. In each circuit, how does the current through the bulb compare to the current through the unknown element?

3. In each circuit, how does the current through the bulb compare to the current through the battery?

C. Predict the relative brightness of bulbs B_1, B_2, and B_3 in the circuits shown. (A dashed box has been drawn around the network of circuit elements that is in series with each of these bulbs.)

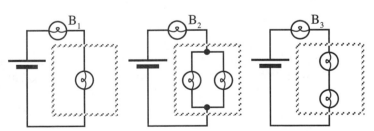

What does your prediction imply about the relative current through each of the *batteries?* Explain.

Have a tutorial instructor show you these circuits so that you can check your answers. Resolve any conflicts between your answers and your observations.

Tutorials in Introductory Physics
McDermott, Shaffer, & P.E.G., U.Wash.

©Prentice Hall
Preliminary Edition, 1998

II. Potential difference

For the remaining circuits in this tutorial use the battery holder with two batteries connected in series. The two-battery combination will be treated as a single circuit element.

A. Set up the circuit with a single bulb and the battery combination as shown. Connect each probe of the voltmeter to a different terminal of the battery holder to measure the potential difference across the battery. Make a similar potential difference measurement across the bulb.

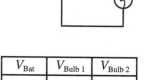

How does the potential difference across the bulb compare to the potential difference across the battery?

V_{Bat}	V_{Bulb}

B. Set up the circuit containing two bulbs in series as shown.

Rank from largest to smallest the currents through bulb 1, bulb 2, and the bulb in the single bulb circuit from part A ($i_{Bulb\,1}$, $i_{Bulb\,2}$, i_{single}). Explain.

Measure the potential difference across each element in the circuit.

V_{Bat}	$V_{Bulb\,1}$	$V_{Bulb\,2}$

1. How does the potential difference across the battery in this circuit compare to the potential difference across the battery in the single-bulb circuit? (See part A.)

2. Rank the potential differences across bulb 1, bulb 2, and the bulb in the single-bulb circuit from part A.

3. How does the potential difference ranking compare to the brightness ranking of the bulbs?

C. Predict what the voltmeter would read if it were connected to measure the potential difference across the network of bulb 1 and bulb 2 together. Explain.

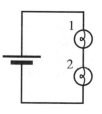

Test your prediction.

How does the potential difference across the network of bulbs compare to the potential difference across the battery?

Tutorials in Introductory Physics
McDermott, Shaffer, & P.E.G., U.Wash.

©Prentice Hall
Preliminary Edition, 1998

D. Set up the circuit with two bulbs in parallel as shown.

Rank the currents through bulb 1, bulb 2, and the bulb in the single bulb circuit from part A. Explain.

How does the current through bulb 1 compare to the current through the battery? Explain.

Measure the potential difference across each circuit element.

V_{Bat}	$V_{Bulb\ 1}$	$V_{Bulb\ 2}$

1. How does the potential difference across the battery in this circuit compare to the potential difference across the battery in the single-bulb circuit?

2. Rank the potential difference across bulb 1, bulb 2, and the bulb in the single bulb circuit from part A.

3. How does the ranking by potential difference compare to the ranking by brightness?

E. Answer the following questions based on the measurements you have made so far.

1. Does the *current through the battery* depend on the circuit to which it is connected? Explain.

2. Does the *potential difference across the battery* depend on the circuit to which it is connected? Explain.

III. Extending the model

Our model for electric circuits includes the idea that, for identical bulbs, the brightness of a bulb is an indicator of the current through the bulb. Based on our observations in this tutorial, we can extend the model to include the idea that, for circuits containing identical bulbs, the brightness of a bulb is also an indicator of the potential difference across the bulb.

A. Set up the circuit with three bulbs as shown and observe their brightness.

Before making the voltmeter measurements, predict the ranking of the potential difference across the battery and each bulb (V_{Bat}, $V_{Bulb\ 1}$, $V_{Bulb\ 2}$, and $V_{Bulb\ 3}$). Explain your prediction.

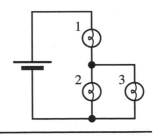

Measure the potential difference across each element in the circuit. If your measurements are not consistent with your ranking above, resolve the inconsistencies.

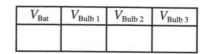

V_{Bat}	$V_{Bulb\ 1}$	$V_{Bulb\ 2}$	$V_{Bulb\ 3}$

B. Before setting up the circuit shown at right:

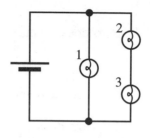

1. Predict the ranking of the currents through the battery and each bulb (i_{Bat}, $i_{Bulb\ 1}$, $i_{Bulb\ 2}$, and $i_{Bulb\ 3}$). Explain.

2. Predict the voltmeter measurements across each of the elements in the circuit shown. Explain.

Prediction:

V_{Bat}	$V_{Bulb\ 1}$	$V_{Bulb\ 2}$	$V_{Bulb\ 3}$

3. Set up the circuit and check your predictions. If your observations and measurements are not consistent with your predictions, resolve the inconsistencies.

Measurement:

V_{Bat}	$V_{Bulb\ 1}$	$V_{Bulb\ 2}$	$V_{Bulb\ 3}$

C. Both circuits at right have more than one path for the current. Sketch all possible current loops on the diagrams. (A "current loop" is *a single path* of conductors that connects one side of the battery to the other.)

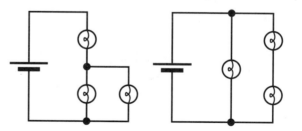

For each of the current loops you have drawn, calculate the sum of the potential differences across the bulbs in that loop. (Use the measurements you made above.)

How do the sums of the potential differences across the bulbs in each loop compare to the potential difference across the battery?

⇨ Check your answer with a tutorial instructor.

Tutorials in Introductory Physics
McDermott, Shaffer, & P.E.G., U.Wash.

©Prentice Hall
Preliminary Edition, 1998

Throughout this tutorial, when you are asked to *predict* the behavior of a circuit, do so *before* setting up the circuit.

I.   Simple RC circuits

A. A capacitor is connected to a battery, bulb and switch as shown. Assume the switch has been closed for an *extended period of time*.

1. *Predict* whether the brightness of the bulb is *the same as, greater than* or *less than* the brightness of a single bulb connected to a battery. Explain.

2. *Predict* how the potential difference across the battery compares to the potential difference across the capacitor plates and to the potential difference across the bulb. Explain.

3. Briefly describe the distribution of charge, if any, on the capacitor plates.

Recall the relationship between the charge on a capacitor and the potential difference across the capacitor. Use this relationship to describe how you could use a voltmeter to determine the charge on a capacitor.

4. Obtain the circuit and a voltmeter. Check your predictions for parts 1 and 2.

Tutorials in Introductory Physics
McDermott, Shaffer, & P.E.G., U.Wash.

©Prentice Hall
Preliminary Edition, 1998

B. Remove the capacitor and the bulb from the circuit.

 1. *Predict* the potential difference across the bulb and the potential difference across the capacitor while these elements are disconnected from the circuit and from each other. Explain.

   Check your prediction.

 2. *Predict* whether the potential difference across the capacitor will *increase, decrease* or *remain the same* if a wire is connected from "ground" to one or the other of the terminals of the capacitor. Explain your reasoning.

   Check your prediction. (You can use a wire with clip leads connected to a metal table leg as a "ground.")

 3. Devise and carry out a method to reduce the potential difference across the capacitor to zero. (This is sometimes called *discharging* the capacitor.)

 4. The capacitor in part A is said to be *charged* by the battery.

   Does the capacitor have a *net* charge after being connected to the battery?

   In light of your answer above, what is meant by *the charge* on a capacitor?

Tutorials in Introductory Physics
McDermott, Shaffer, & P.E.G., U.Wash.

©Prentice Hall
Preliminary Edition, 1998

II. Charging and discharging capacitors

A. Suppose an uncharged capacitor is connected in series with a battery and bulb as shown.

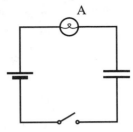

1. *Predict* the behavior of the bulb when the switch is closed. Explain.

   Set up the circuit and check your prediction. If your prediction is in conflict with your observation, how can you account for your observation?

2. *Without using a voltmeter,* determine the potential difference across the capacitor at the following times:

   • *just after* the switch is closed. Explain how you can tell. (*Hint:* Compare the brightness of the bulb to the brightness of a bulb connected to a battery in a single bulb circuit without a capacitor.)

   • *a long time after* the switch is closed. Explain how you can tell.

   Use a voltmeter to check your predictions. (*Hint:* Be sure to discharge the capacitor completely after each observation.)

B. Suppose that instead of connecting the uncharged capacitor to the single bulb A, you connected it to the two-bulb circuit shown at right.

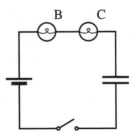

1. *Predict* how the *initial brightness* of bulb B compares to the initial brightness of bulb C. Explain.

2. *Predict* how the *initial brightness* of bulb B compares to the initial brightness of bulb A above. Explain.

   Discharge the capacitor and then set up the circuit with the uncharged capacitor and check your predictions. If your prediction is in conflict with your observation, how can you account for your observation?

3. *Predict* how the final charge on the capacitor compares to the final charge on the capacitor from part A.

Use a voltmeter to check your prediction.

C. Suppose that the bulbs were connected in parallel, rather than in series.

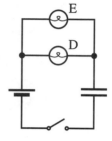

1. *Predict* how the *initial brightness* of bulb D compares to the initial brightness of bulb E. Explain.

2. *Predict* how the *initial brightness* of bulb D compares to the initial brightness of bulbs A, B, and C above. Explain.

3. *Predict* how the final charge on the capacitor compares to the final charge on the capacitor from part A. Explain.

Set up the circuit and check your predictions. If your prediction is in conflict with your observation, how can you account for your observation?

D. After completing the exercises above, two students make the following comments:

Student 1: *"The capacitor with two bulbs in series got charged up a lot more than the capacitor with two bulbs connected in parallel because the series circuit charged the capacitor for a longer period of time."*

Student 2: *"I disagree, the bulbs in the parallel circuit were brighter so this capacitor gained more charge."*

Do you agree with student 1, student 2, or neither? Explain your reasoning.

Tutorials in Introductory Physics
McDermott, Shaffer, & P.E.G., U.Wash.

©Prentice Hall
Preliminary Edition, 1998

E. Suppose that a different capacitor of smaller capacitance were connected to the battery and a single bulb in series.

   1. *Predict* how the initial potential difference across the bulb compares to the initial potential difference across the bulb in part A.

   2. *Predict* how the initial brightness of the bulb compares to the brightness of the single bulb in part A. Explain.

   3. *Predict* how the final amount of charge on the capacitor would compare to the final amount of charge on the capacitor from part A.

       Set up the circuit and check your predictions. If your prediction is in conflict with your observation, how can you account for your observation?

III. Multiple capacitors

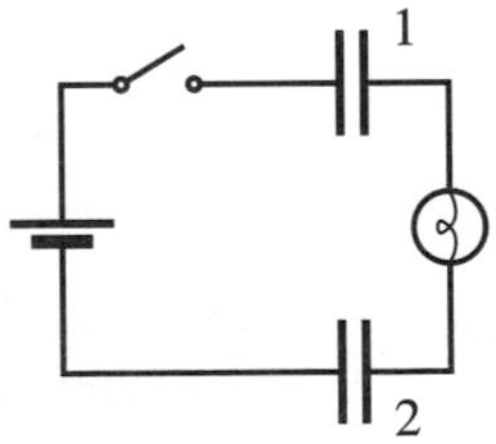

A bulb is connected to a battery and two capacitors as shown at right. Suppose that C_1 is less than C_2.

A. Before connecting the circuit a student makes the following prediction:

> *"Current flows from the positive side of the battery to the negative side of the battery. Since the bulb is isolated from the battery on both sides by the capacitors, the bulb will not light."*

Do you agree or disagree with this prediction? Explain.

       Obtain a second capacitor from your instructors and check your prediction.

B. Make the following predictions on the basis of your observations of this circuit. Do not use a voltmeter.

1. Just after the switch is closed:

- What is the potential difference across the bulb? Explain how you can tell from the brightness of the bulb.

- What is the potential difference across each of the capacitors? Explain your reasoning.

2. A long time after the switch is closed:

- What is the potential difference across the bulb? Explain how you can tell.

- What is the sum of the potential differences across the two capacitors? Explain.

- Is the potential difference across capacitor 1 *greater than*, *less than*, or *equal to* the potential difference across capacitor 2? Explain.

- Is the final charge on capacitor 1 *greater than*, *less than*, or *equal to* the final charge on capacitor 2? Explain.

Use the voltmeter to check your predictions in part B.

Tutorials in Introductory Physics
McDermott, Shaffer, & P.E.G., U.Wash.

©Prentice Hall
Preliminary Edition, 1998

Magnetism

Tutorials in Introductory Physics
McDermott, Shaffer, & P.E.G., U.Wash.

I.   Magnetic materials

A. Investigate the objects that you have been given (magnets, metals, cork, plastic, wood, etc.).
    Separate the objects into three classes based on their interactions with each other.

1. List the objects in each of your classes.

<u>class 1</u>   <u>class 2</u>   <u>class 3</u>

2. Fill out the table below with a word or two describing the interaction between members
    of the same and different classes.

Table of Interactions

	class 1	class 2	class 3
class 1			
class 2			
class 3			

3. Are all metals in the same class?

4. To which class do magnets belong? Are all the objects in this class magnets?

B. Obtain a permanent magnet and an object that is attracted to the magnet but not repelled.
    Imagine that you do not know which object is the magnet. Using only these two objects, find
    a way to determine which object is the permanent magnet. (*Hint:* Are there parts on either
    object that do not interact as strongly as other parts?)

C. The parts of a permanent magnet that interact most strongly with other materials are called
    the *poles* of a magnet.

    How many magnetic poles does each of your magnets have? Explain how you found them.

    How many different *types* of poles do you have evidence for so far? Explain.

    Using three magnets, find a way to distinguish one *type* of pole from another.

Tutorials in Introductory Physics
McDermott, Shaffer, & P.E.G., U.Wash.

©Prentice Hall
Preliminary Edition, 1998

D. Describe how an *uncharged* pith ball suspended from a string can be used to test whether an object is charged.

Predict what will happen when an uncharged pith ball is brought near one of the poles of the magnet. Explain.

Obtain a pith ball and test your prediction. Record your results.

Based on your observations above, *predict* what will happen when the pith ball is brought near the other pole of the magnet. Test your prediction.

Is there a *net charge* on the north (or south) pole of a magnet? Explain.

E. A paper clip is attached to a string and suspended from a straw. It is then placed so that it hangs inside an aluminum-foil lined cup as shown.

1. *Predict* what will happen to the paper clip when a charged rod is brought near the cup. Explain in terms of the electric field inside the foil-lined cup.

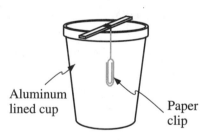

Aluminum
lined cup

Paper
clip

Obtain the equipment and test your prediction. Discuss this experiment with your partners.

Predict what you would observe if the paper clip were outside the cup. Explain your reasoning, then check your prediction

2. Bring a magnet near the cup and observe what happens to the paper clip inside the cup. Record your observations.

F. Based on your observations in parts D and E above, would you say that a magnetic interaction is the *same as* or *different from* an electrical interaction? Explain.

Tutorials in Introductory Physics
McDermott, Shaffer, & P.E.G., U.Wash.

©Prentice Hall
Preliminary Edition, 1998

II.   Magnetic fields

We have observed that magnets interact even when they are not in direct contact. In electrostatics we used the idea of an electric field to account for the interaction between charges that were separated from one another. With magnetic interactions, we similarly define a *magnetic field*.

A. Obtain a compass from a tutorial instructor.

1.   Use the compass to explore the region around a bar magnet.

Describe the behavior of the compass needle both near the poles of the magnet and in the region between the poles.

To which class of objects from section I does the compass needle belong? Explain.

2.   Move the compass far away from all other objects. Shake the compass and describe the behavior of the compass needle.

Does the needle behave as if it is in a magnetic field?

We can account for the behavior of the compass needle by supposing that it interacts with the earth and that the earth belongs to one of the categories from section I.

To which class of objects from section I do your observations suggest the earth belongs? Explain how you can tell.

3.   We define the *north pole* of a magnet as the end that points toward the arctic region of the earth when the magnet is free to rotate and is not interacting with other nearby objects.

On the basis of this definition, is the geographic north pole of the earth a magnetic north pole or a magnetic south pole?

Use your compass to identify the north pole of an unmarked bar magnet.

B. Place a bar magnet on an enlargement of the diagram at right.

A.

S ┃ N

.B

.C

1. Place the compass at each of the lettered points on the enlargement and draw an arrow to show the direction in which the north end of the compass points.

Discuss with your partners how the interaction of the compass with the magnet depends on the distance from the bar magnet and the location around the bar magnet.

E.

D.

Devise a method by which you can determine the approximate relative magnitudes of the magnetic field at each of the marked locations. Explain your reasoning.

2. We *define* the direction of the magnetic field at a point as the direction in which the north end of a compass needle points when the compass is placed at that point.

Make the arrows on your enlargement into magnetic field *vectors* (*i.e.,* draw them so that they include information about both the magnitude and direction of the field).

C. Obtain some small magnets and stack them north-to-south until you have a bar about the same length as your bar magnet. Place them on an enlargement of the diagram at right.

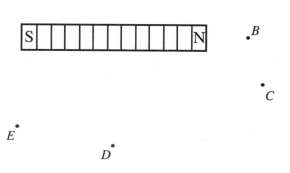

1. On the enlargement, sketch the magnetic field *vectors* at the locations *A-E.*

How does the magnetic field of the stack of magnets compare to the magnetic field of the bar magnet?

2. Break the stack in half and investigate the breaking points. Describe how many north and how many south poles result.

What does your observation suggest about how a bar magnet would behave when broken in half?

Tutorials in Introductory Physics
McDermott, Shaffer, & P.E.G., U.Wash.

©Prentice Hall
Preliminary Edition, 1998

3. On your enlargement draw the magnetic field vectors at the six locations *A-F* when just the right half of the stack of magnets is present.

    Using a different color pen, draw the magnetic field vectors when just the left half of the stack of magnets is present.

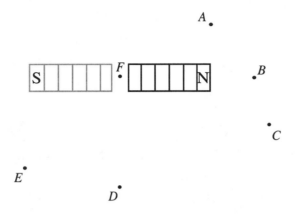

    Compare the field vectors for the two half-stacks of magnets to the field vectors for the whole stack.

    Is your observation consistent with the idea that magnetic fields obey the principle of superposition? Explain.

    From your observations, what can you infer about the direction of the magnetic field inside a bar magnet? Explain.

    Sketch magnetic field vectors for a few points inside the magnet.

    Does the magnetic field of a bar magnet always point away from the north pole and toward the south pole of the magnet? Explain.

    What can you infer about the strength of the magnetic field inside the magnet as compared to outside the magnet?

I.   The magnetic force on a current-carrying wire in a magnetic field

Obtain the following equipment:

- magnet
- wooden dowel
- ring stand and clamp
- battery and case
- switch
- two paper clips
- two alligator-clip leads
- 30 cm piece of connecting wire
- magnetic compass
- enlargement showing magnet and wire

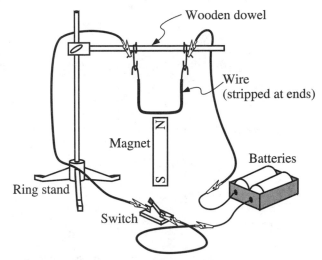

Hang the connecting wire from the paper clips as shown so that it swings freely. Do not close the switch until told to do so.

A. On an enlargement of the figure below, sketch field lines representing the magnetic field of the bar magnet. Show the field both inside and outside the magnet.

On the diagram, indicate the direction of the current through the wire when the switch is closed.

Predict the direction of the force exerted on the wire by the magnet. Explain your reasoning.

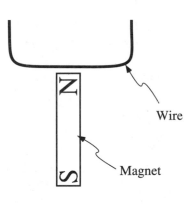

Check your prediction. (Do not leave the switch closed for more than a few seconds. The battery and wires will become hot if the circuit is connected for too long.)

B. Make predictions for the following five situations based on what you observed in part A. Check your answers only after you have made all five predictions.

1. The magnet is turned so that the south pole is near the wire while the switch is closed.

    Prediction:

    Observation:

2. The leads to the battery are reversed (consider both orientations of the magnet).

    Prediction:

    Observation:

Tutorials in Introductory Physics
McDermott, Shaffer, & P.E.G., U.Wash.

©Prentice Hall
Preliminary Edition, 1998

3. The north pole of the magnet is held near the wire but the switch remains open.

   Prediction:

   Observation:

4. The north pole of the magnet is held: (a) closer to the wire and (b) farther from the wire.

   Prediction:

   Observation:

5. The magnet is turned so that it is parallel to the wire while the switch is closed.

   Prediction:

   Observation:

Resolve any discrepancies between your predictions and your observations. (*Hint:* Consider the *vector* equation for the magnetic force on a current carrying wire in a magnetic field: $\vec{F} = i\vec{L} \times \vec{B}$.)

II. The magnetic field of a current-carrying wire

A. Suppose you place a small magnet in a magnetic field and allow it to rotate freely. What will the orientation of the small magnet be relative to the external magnetic field lines? Draw a sketch to illustrate your answer.

B. Suppose you hold a magnetic compass near a current-carrying wire as shown. (A *magnetic compass* is a magnet that can rotate freely.) The face of the compass is parallel to the tabletop.

   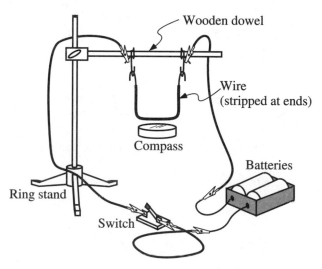

   1. *Predict* the orientation of the compass needle when the switch is closed. Sketch a diagram that shows the wire, the direction of the current through it, the direction of the magnetic field directly below the wire, and the predicted orientation of the compass needle.

   2. Check your answer. If the deflection of the needle is not what you predicted, resolve the discrepancy. (*Hint:* Is there more than one magnetic field affecting the compass?)

Tutorials in Introductory Physics
McDermott, Shaffer, & P.E.G., U.Wash.

©Prentice Hall
Preliminary Edition, 1998

C. Now suppose that you hold the compass at some other locations near the wire (*e.g.,* directly above the wire or to one side of a vertical wire). For each location, *predict* the orientation of the compass needle when the switch is closed. Make sketches to illustrate your predictions.

Check your answers. If the orientation of the compass needle is not what you predicted, resolve the discrepancy.

D. Sketch the magnetic field lines of a current-carrying wire. Include the direction of the current in the wire in your sketch.

III. Current loops and solenoids

A. A wire is formed into a loop and the leads are twisted together. The sides of the loop arc labcled A–D. The direction of the current is shown. (The diagram uses the convention that ⊙ indicates current out of the page and ⊗ indicates current into the page.)

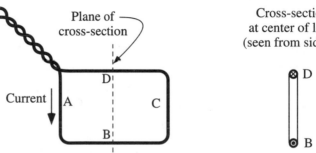

1. On the top two diagrams at right, sketch magnetic field lines for the loop. Base your answer on your knowledge of the magnetic field of a current-carrying wire.

Explain why it is reasonable to ignore the effect of the magnetic field from the wire leads.

2. Consider the magnetic field of a bar magnet.

How are the magnetic field lines for the current loop similar to those for a short bar magnet?

Can you identify a "north" and a "south" pole for a current loop?

Devise a rule by which you can use your right hand to identify the magnetic poles of the loop from your knowledge of the direction of the current.

B. A small current loop is placed near the end of a large magnet as shown.

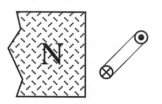

1. Draw vectors to show the magnetic force on each side of the loop.

    What is the net effect of the magnetic forces exerted on the loop?

2. Suppose that the loop were to rotate until oriented as shown.

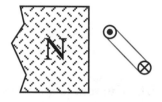

    Now, what is the net effect of the magnetic forces exerted on the loop?

    Is there an orientation for which there is no net torque on the loop? Draw a diagram to illustrate your answer.

3. Are your results above consistent with regarding the current loop as a small magnet? Label the poles of the current loop in the diagrams above and check your answer.

Tutorials in Introductory Physics
McDermott, Shaffer, & P.E.G., U.Wash.

©Prentice Hall
Preliminary Edition, 1998

C. A solenoid is an arrangement of many current loops placed together as shown below. The current through each loop is the same and is in the direction shown. Obtain or draw an enlarged copy of the figure.

1. Use the principle of superposition to draw on the enlarged copy a vector at each of the points *A–E* to indicate the direction *and* the relative strength of the magnetic field.

2. Sketch the magnetic field lines on the enlarged copy.

    Describe the magnetic field near the center of the solenoid.

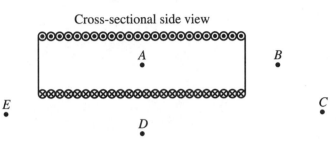

Cross-sectional side view

3. How does the field of the solenoid at points *A–E* compare with that of a bar magnet (both inside and outside)?

    Which end of the solenoid corresponds to a north pole? Which end corresponds to a south pole?

4. How would the magnetic field at any point within the solenoid be affected by the following changes? Explain your reasoning in each case.

    • The current through each coil of the solenoid is increased by a factor of two.

    • The number of coils in each unit length of the solenoid is increased by a factor of two, with the current through each coil remaining the same.

Electromagnetism

I. Induced currents

A. A copper wire loop is placed in a uniform magnetic field as shown. Determine whether there would be a current through the wire of the loop in each case below. Explain your answer in terms of magnetic forces exerted on the charges in the wire of the loop.

- The loop is stationary.

- The loop is moving to the right.

- The loop is moving to the left.

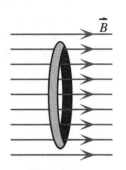

B. Suppose that the loop is now placed in the magnetic field of a solenoid as shown.

1. Determine whether there would be a current through the wire of the loop in each case below. If so, give the direction of the current. Explain in terms of magnetic forces exerted on the charges in the wire of the loop.

   - The loop is stationary.

   - The loop is moving toward the solenoid.

   - The loop is moving away from the solenoid.

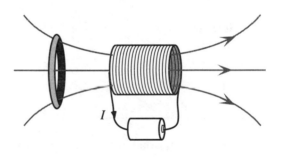

2. For each case in which there is an induced current, determine:

   - the direction of the *magnetic moment* of the loop. (*Hint:* Find the direction of the magnetic field at the center of the loop due to the induced current in the loop. The magnetic moment is a vector that points in this same direction.)

   - whether the loop is *attracted toward* or *repelled from* the solenoid.

   - whether the force exerted on the loop tends to *increase* or to *decrease* the relative motion of the loop and solenoid.

Tutorials in Introductory Physics
McDermott, Shaffer, & P.E.G., U.Wash.

©Prentice Hall
Preliminary Edition, 1998

C. In each of the diagrams below, the position of a loop is shown at two times, t_1 and t_2 $(t_1 < t_2)$. The loop starts from rest in each case and is displaced to the right in Case A and to the left in Case B. On the diagrams indicate:

- the direction of the induced current through the wire of the loop,
- the magnetic moment of the loop,
- an area vector for each loop,
- the sign of the flux due to the external magnetic field (at both times t_1 and t_2), and
- the sign of the induced flux (at both times t_1 and t_2).

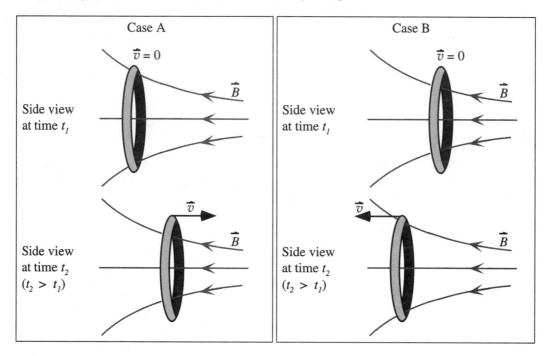

D. State whether you agree or disagree with each of the students below. If you agree, explain why. If you disagree, cite a specific case for which the student's statement does not give the correct answer. (*Hint:* Consider cases A and B above.)

Student 1: *"The magnetic field due to the loop always opposes the external magnetic field."*

Student 2: *"The flux due to the loop always has the opposite sign as the flux due to the external magnetic field."*

Student 3: *"The flux due to the loop always opposes the change in the flux due to the external magnetic field."*

⇨ Before continuing, check your answer with a tutorial instructor.

Tutorials in Introductory Physics
McDermott, Shaffer, & P.E.G., U.Wash.

©Prentice Hall
Preliminary Edition, 1998

II. Lenz' law

A. The diagram at right shows a copper wire loop in a uniform magnetic field. The magnitude of the field is *decreasing* with time.

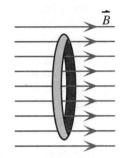

1. Would you predict that there would be a current through the loop:

   • if you were to use the idea that there is a magnetic force exerted on a charge moving in a magnetic field? Explain your reasoning.

   • if you were to use the reasoning of the student in part D of section I with whom you agreed? Explain.

2. It is *observed* that there is an induced current through the wire loop in this case. Use the appropriate reasoning above to find the direction of the current through the wire of the loop.

To understand the interaction between the wire loops and solenoids in section I, we can use the idea that a force is exerted on a charged particle moving in a magnetic field. In each of those cases there was an induced current when there was relative motion between the solenoid and the wire loop. In other situations such as the one above, however, there is an induced current in the wire loop even though there is no relative motion between the wire loop and the solenoid. There is a general rule called *Lenz' law* that we can use in *all* cases to predict the direction of the induced current.

B. Discuss the statement of Lenz' law in your textbook with your partners. Make sure you understand how it is related to the statement by the student with whom you agreed in part D of section I.

C. A wire loop moves from a region with no magnetic field into a region with a uniform magnetic field pointing into the page.

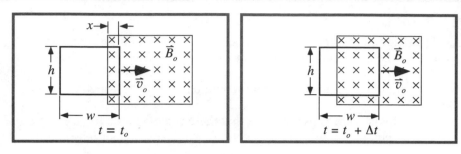

The loop is shown at two instants in time, $t = t_o$ and $t = t_o + \Delta t$.

1. Is the magnetic flux through the loop due to the external field *positive, negative,* or *zero:*

   a. at $t = t_o$?

   b. at $t = t_o + \Delta t$?

2. Is the *change* in flux due to the external field in the interval Δt *positive, negative,* or *zero?*

3. Use Lenz' law to determine the sign of the flux due to the induced current in the loop.

4. What is the direction of the current in the loop during this time interval?

D. At two later instants, $t = t_1$ and $t = t_1 + \Delta t$, the loop is located as shown.

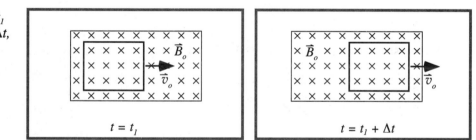

1. Use Lenz' law to determine the sign of the flux due to the current induced in the loop. Explain.

2. Describe the current in the loop during this time interval.

3. Consider the following student dialogue:

   Student 1: *"The sign of the flux is the same as it was in part C. So the current here will also be counter-clockwise."*

   Student 2: *"I agree. If I think about the force on a positive charge on the leading edge of the loop, it points towards the top of the page. That's consistent with a counter-clockwise current."*

   Do you agree with either student? Explain.

Tutorials in Introductory Physics
McDermott, Shaffer, & P.E.G., U.Wash.

©Prentice Hall
Preliminary Edition, 1998

I.   Faraday's law

Two loops of the same radius are held near a solenoid.
Both loops are the same distance from the end of the
solenoid and are the same distance from the axis of the
solenoid.

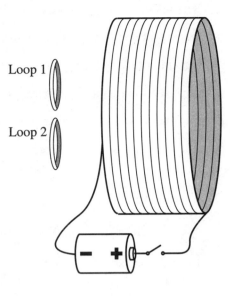

A.   The resistance of loop 2 is greater than that of loop 1.
(The loops are made from different materials.)

1.   Is there a current induced through the wire of
either of the loops:

• before the switch is closed? Explain.

• just after the switch is closed? Explain.

• a long time after the switch is closed? Explain.

2.   For the period of time that there is a current induced through the wire of the loops, find
the direction of the current.

3.   The ratio of the induced currents for the two loops is found by experiment to be equal to
the inverse of the ratio of the resistances of the loops.

What does this observation imply about the ratio of the induced emfs in the two loops?

B.   Suppose that loop 2 were replaced by a wooden loop.

• Would there still be an emf in the loop?

• Would there still be a current induced in the wood loop?

C.   Suppose that loop 2 were removed completely. Consider the circular path that the wire of
loop 2 used to occupy.

• Would there still be an emf along the path? Explain.

• Would there still be a current along the path? Explain.

Tutorials in Introductory Physics
McDermott, Shaffer, & P.E.G., U.Wash.

©Prentice Hall
Preliminary Edition, 1998

The results of the previous exercises are consistent with the idea that a change in the magnetic flux through the surface of a loop results in an *emf* in that loop. If there is a conducting path around the loop (*e.g.*, a wire), there will be a current. The emf is independent of the material of which the loop is made; the current is not. It is found by experiment that the induced emf is proportional to the rate of change of the magnetic flux through the loop. This relationship is called *Faraday's law*. The direction of any induced current is given by Lenz' law.

D. Three loops, all made of the same type of wire, are placed near the ends of identical solenoids as shown. The solenoids are connected in series. Assume that the magnetic field near the end of each of the solenoids is uniform.

Loop 2 consists of two turns of a single wire that is twice as long as the wire used to make loop 1. Loop 3 is made of a single wire that is half as long as the wire used to make loop 1.

Just after the switch has been closed, the current through the battery begins to increase. The following questions concern the period of time during which the current is increasing.

1. Let $\mathcal{E}$ represent the induced emf of loop 1. Find the induced emf in each of the other loops in terms of $\mathcal{E}$. Explain your reasoning.

Loop 1

Single loop of radius *r*

2. Let *R* represent the resistance of loop 1. Find the resistance of each of the other loops in terms of *R*. Explain.

Loop 2

Double loop of radius *r* made from a single wire

3. Find the current induced through the wire of each of the loops in terms of $\mathcal{E}$ and *R*.

Loop 3

Single loop of radius *r*/2

Tutorials in Introductory Physics
McDermott, Shaffer, & P.E.G., U.Wash.

©Prentice Hall
Preliminary Edition, 1998

II. Applications

A. Galvanometer

Obtain a device like the one shown below. It contains a coil made of many loops of wire and a magnet suspended so that it is free to swing. A pointer has been attached to the magnet so that a small swing of the magnet will result in a large deflection of the pointer. When there is no current through the coil, the magnet is horizontal and the pointer is vertical.

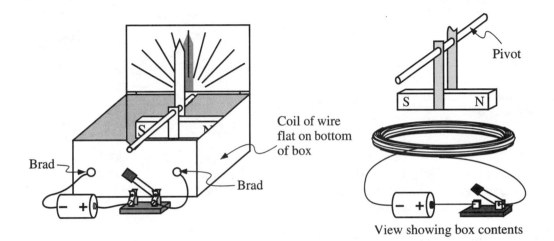

View showing box contents

Predict the deflection of the pointer (if any) when the switch is closed. Explain the reasoning you used to make your prediction.

Connect the circuit and observe the deflection of the pointer. If your observation is in conflict with your prediction, discuss your reasoning with a tutorial instructor.

The device above is called a *galvanometer* and can be used to detect current. If the scale on the galvanometer has been calibrated to measure amperes, the device is called an *ammeter*.

B. Simple electric motor

Obtain the equipment illustrated at
right and assemble it as shown.

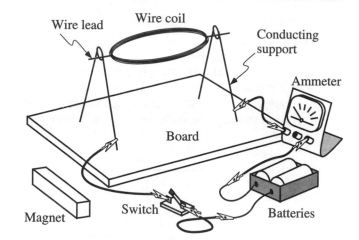

You should have:

- a magnet, a battery, a switch,
  some connecting wire, and an
  ammeter.

- a copper wire coil. The ends of
  the wire leads to the coil have
  been stripped of the insulating
  enamel coating so that half the
  wire is bare.

- two conducting supports for the leads to the coil.

1. Examine the leads to the wire coil closely, so that you
    understand which portion of the wire has been stripped
    of the insulating coating.

    For what orientations of the coil will there be a current
    through it due to the battery?

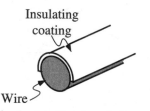

Wire lead half stripped of enamel

Check your answer by closing the switch and observing the deflection of the ammeter as
you rotate the coil manually through one complete revolution.

2. Hold one pole of the magnet near the coil. Close the switch. If the coil does not begin to
    spin, adjust the location of the magnet or gently rotate the coil to start it spinning.

    Use the ideas that we have developed in this and previous tutorials to explain the motion
    of the wire coil. (The questions that follow may serve as a guide to help you develop an
    understanding of the operation of the motor.)

Tutorials in Introductory Physics
McDermott, Shaffer, & P.E.G., U.Wash.

©Prentice Hall
Preliminary Edition, 1998

3. When the coil is in the position shown, there is a current, *I*, through it.

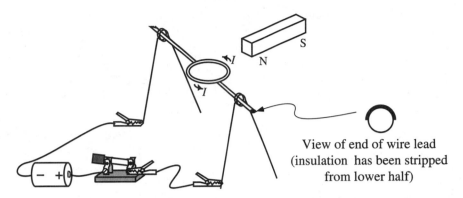

View of end of wire lead
(insulation has been stripped
from lower half)

a. The coil is manually started spinning so that it rotates clockwise.

During which portions of the cycle does the coil form a complete circuit with the battery such that there is a current through the wire of the coil?

The current results in a magnetic moment that interacts with the magnetic field of the magnet. Will the interaction tend to *increase* or to *decrease* the angular speed of the coil? Explain.

b. The coil is manually started spinning so that it rotates counterclockwise:

During which portions of the cycle does the coil form a complete circuit with the battery so that there is a current through the wire of the coil?

The current results in a magnetic moment that interacts with the magnetic field of the magnet. Will the interaction tend to *increase* or to *decrease* the angular speed of the coil? Explain.

Check that the behavior of your motor is consistent with your answers.

4. Consider the following questions about the motor:

• Why was insulated wire used for the coil? Would bare wire also work? Explain.

• Would you expect the motor to work if the leads to the coil were stripped completely? Explain.

5. Predict the effect on the motor of (i) reversing the leads to the battery and (ii) reversing the orientation of the magnet.

Check your predictions.

C. Electric generator

Remove the battery and ammeter from the circuit in part B and insert a μ-ammeter as shown.

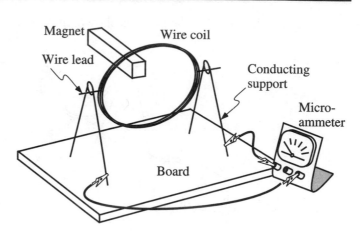

1. Suppose that the coil is made to spin by an external agent such as yourself.

    Predict the deflection of the μ-ammeter needle throughout a complete revolution of the coil.

    How would your prediction change if:

    • the coil were made to rotate the other way?

    • the poles of the magnet were reversed?

2. Check your predictions by gently rotating the coil so that it spins for a little time on its own before coming to a stop.

When the coil of the apparatus above is made to spin by an external agent, the apparatus is called an *electric generator*.

Tutorials in Introductory Physics
McDermott, Shaffer, & P.E.G., U.Wash.

©Prentice Hall
Preliminary Edition, 1998

Waves

I. Pulses on a spring

A tutorial instructor will demonstrate various pulses on a stretched spring. Observe the motion of the pulse and of the spring in each case and discuss your observations with your classmates.

A. A piece of yarn has been attached to the spring. How did the motion of the yarn compare to the motion of the pulse for each type of pulse that you observed?

The terms *transverse* or *longitudinal* are often used to describe the types of pulses you have observed in the demonstration. To what feature of a pulse do these terms refer?

For the rest of this tutorial we will focus on *transverse* pulses along the spring.

B. During the demonstration, did any of the following features change *significantly* as the pulse moved along the spring? (Ignore what happened when the pulse reached the end of the spring.)

- the amplitude of the pulse

- the width of the pulse

- the shape of the pulse

- the speed of the pulse

C. During the demonstration, each of the following quantities was changed. Did any of the changes significantly affect the speed of the pulse? If so, how?

- the tension (*e.g.,* by stretching the spring to a greater length)

- the amplitude of the pulse

- the width of the pulse

- the shape of the pulse

Tutorials in Introductory Physics
McDermott, Shaffer, & P.E.G., U.Wash.

©Prentice Hall
Preliminary Edition, 1998

II. Superposition

A. The snapshots below show two pulses approaching each other on a spring. The pictures were taken at equal time intervals. The pulses are on the "same side" of the spring (*i.e.,* each displaces the spring toward the top of the page).

1. When the pulses meet, does each pulse continue to move in the direction it was originally moving, or does each reverse direction?

    Give evidence from the photos to support your answer.

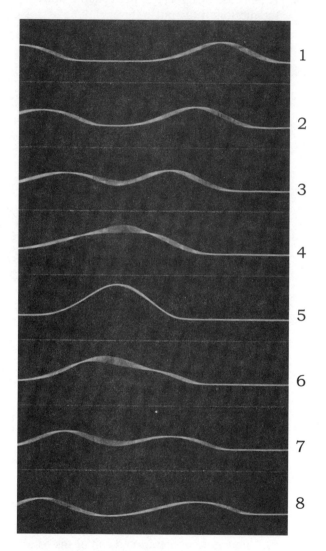

2. When the pulses completely overlap, as shown in snapshot 5, how does the shape of the disturbance in the spring compare to the shapes of the individual pulses?

3. You have used the *principle of superposition* in previous physics classes.

    Describe how you could use this principle to determine the shape of the spring at any instant while the pulses "overlap."

4. Two pulses (1 and 2) approach one another as shown. The bottom diagram shows the location of pulse 1 a short time later.

    In the space at right, sketch the location of pulse 2 at this later time. On the same diagram, sketch the shape of the spring at this instant in time.

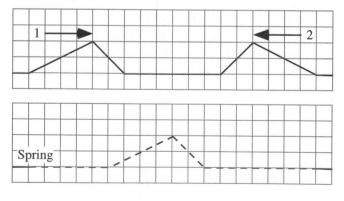

Tutorials in Introductory Physics
McDermott, Shaffer, & P.E.G., U.Wash.

©Prentice Hall
Preliminary Edition, 1998

B. Two pulses of equal width and equal amplitude approach each other on *opposite sides* of a spring *(i.e.,* the pulses displace the spring in opposite directions). The snapshots below were taken at equal time intervals.

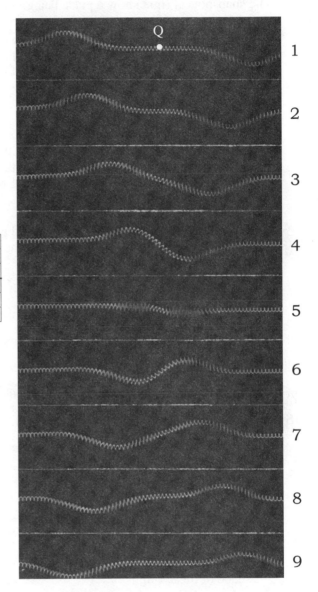

1. Is the behavior of the spring consistent with the principle of superposition? If so, which quantity is "added" in this case? If not, explain why not.

2. Below is a simplified representation of both individual pulses at a time between the instants shown in snapshots 4 and 5.

    Sketch the shape of the spring at the instant shown.

3. Let point *Q* be the point on the spring midway between the pulses.

    Describe the motion of point *Q* during the time interval shown.

4. Which, if any, of the following changes would affect the motion of point *Q*? Explain.

    • doubling the amplitude of both pulses

    • doubling the amplitude of just one pulse

    • doubling the width of just one pulse

5. Consider an *asymmetric* pulse as shown.

    What shape would a second pulse need to have in order that point *Q* not move as the two pulses pass each other?

    On the diagram, indicate the shape, location, and direction of motion of the second pulse at the instant shown.

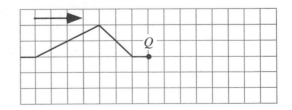

©Prentice Hall
Preliminary Edition, 1998

III. Reflection

A. Reflection from a fixed end

The snapshots at right show a pulse being *reflected* from the end of a spring that is held fixed in place.

1. Describe the similarities and differences between the *incident pulse* (the pulse moving toward the fixed end) and the *reflected pulse*.

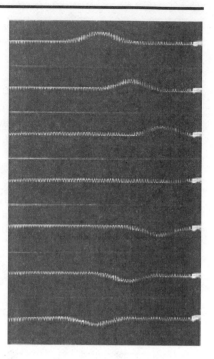

2. Consider the situation in part B of section II, in which two pulses on opposite sides of a spring meet. Use a piece of paper to cover the right half of those photographs so that the portion of spring *to the left of point Q* is uncovered.

    How does the behavior of the uncovered portion of spring (including point *Q*) compare to the behavior of the spring shown at right?

The results of the exercise above suggest a model for the reflection of pulses from fixed ends of springs. We imagine that the spring extends past the fixed end and that we can send a pulse along the imaginary portion toward the fixed end. We choose the shape, orientation, and location of the imagined pulse so that as it passes the incident pulse, the end of the spring *remains fixed*. (Such a condition that governs the behavior of the end of the spring is an example of a *boundary condition*.) In this case, the reflected and imagined pulses have the same shape and orientation.

3. A pulse with speed 1.0 m/s is incident on the fixed end of a spring.

    Determine the shape of the spring at (a) *t* = 0.2 s, (b) *t* = 0.4 s, and (c) *t* = 0.6 s.

    How does the shape of the reflected pulse compare to that of the incident pulse?

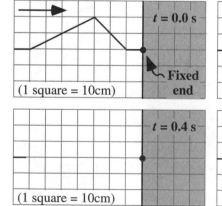

B. Reflection from a free end

Before you leave tutorial, observe a demonstration of a pulse reflecting from the *free end* of a spring. Record your observations. You will investigate this situation in the homework.

Tutorials in Introductory Physics
McDermott, Shaffer, & P.E.G., U.Wash.

©Prentice Hall
Preliminary Edition, 1998

I. Reflection and transmission at a boundary

The photographs below illustrate the behavior of two springs joined end-to-end when a pulse reaches the boundary between the springs. The snapshots were taken at equal time intervals.

A. Describe what happens after the pulse reaches the boundary between the springs.

   Compare the widths of the incident and transmitted pulses.

B. Compare the speed of a pulse in one spring to the speed of a pulse in the other spring. Make this comparison in two ways:

   1. Use the information contained in two or more snapshots. Explain.

   2. Use the information contained in only a single snapshot (*e.g.,* snapshot 8). Explain.

C. In answering the questions below, assume that each spring has approximately uniform tension.

   1. How does the tension in one spring compare to the tension in the other spring? Explain.

   2. How does the linear mass density, μ, of one spring compare to the linear mass density of the other? Explain.

⇨ Check your answers with a tutorial instructor.

Tutorials in Introductory Physics
McDermott, Shaffer, & P.E.G., U.Wash.

©Prentice Hall
Preliminary Edition, 1998

II.   Transmission of multiple pulses

Imagine that two identical pulses are sent toward the boundary between the two springs from section I, as illustrated below. For this part of the tutorial, ignore reflected pulses.

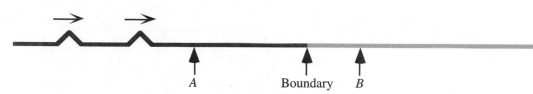

A
Boundary   B

A.   Imagine that you measure the time interval that starts when the crest of the first pulse reaches point *A* and ends when the crest of the second pulse reaches that same point. Also imagine that one of your partners measures the corresponding time interval for the transmitted pulses at point *B*.

Would the time interval for the incident pulses (at point *A*) be *greater than, less than,* or *equal to* the time interval for the transmitted pulses (at point *B*)? (*Hint:* Imagine a third person measuring this time interval at the boundary.)

Would the distance between transmitted crests be *greater than, less than,* or *equal to* the distance between incident crests? Explain.

B.   Is the time it takes a single incident pulse to pass by point *A greater than, less than,* or *equal to* the time it takes a single transmitted pulse to pass by point *B?*

Explain how the change in the width of the pulse as it passes from the first spring to the second is a direct consequence of the difference in speed in the two springs.

On the diagram above, sketch the transmitted pulses showing the widths and spacing of the transmitted pulses relative to the incident pulses.

⇨   Check your answers for parts A and B with a tutorial instructor.

Tutorials in Introductory Physics
McDermott, Shaffer, & P.E.G., U.Wash.

©Prentice Hall
Preliminary Edition, 1998

III. Reflection and transmission at a boundary revisited

The springs in the photograph at right are the same as in the photographs on the first page. However, now a pulse approaches the boundary between the springs from the right.

A. After the trailing edge of the incident pulse has reached the boundary, will there be a reflected pulse?

> *If so:*   On which side of the spring will the reflected pulse be located? How will its width compare to the width of the incident pulse?

> *If not:*   Explain why not.

How will the transmitted pulse compare to the incident pulse?

In the space below the photograph, make a sketch that shows the shape of the springs at an instant after the incident pulse is completely transmitted. Your sketch should illustrate the relative widths of the pulse(s) and their relative distance(s) from the boundary as well as which side of the spring each pulse is on.

B. Ask a tutorial instructor for the time sequence of photographs that illustrates this situation so that you can check your predictions.

If your prediction was incorrect, identify those parts of your prediction that were wrong.

IV. A model for reflection at a boundary

We have observed that reflection occurs when a pulse reaches the boundary between two springs, that is, where there is an abrupt *change* in medium. We would like to be able to predict whether the boundary will act more like a fixed end or more like a free end.

A. In the situation illustrated in section I, are the incident and reflected pulses on the *same* side of the spring, or are they on *opposite* sides of the spring?

On the basis of this observation, does it appear that the reflection at the boundary is more like reflection from a fixed end or a free end?

B. Which of the following quantities are different on the two sides of the boundary?

- tension

- linear mass density

- wave speed

What property or properties of the two media could you use to predict whether the boundary will act more like a fixed end or more like a free end? (It may help to consider limiting cases, *i.e.*, very large or very small values of the properties.)

Describe how you could predict whether the reflected pulse will be on the same side of the spring as the incident pulse or whether it will be on the opposite side.

Describe how you could predict whether the transmitted pulse will be on the same side of the spring as the incident pulse or whether it will be on the opposite side.

C. Imagine that a pulse on a spring is approaching a boundary. Would the boundary act more like a fixed end or more like a free end if the spring is connected to:

- a very massive chain?

- a very light fishing line?

I.   Water waves passing from shallow water to deep water

A. The diagram at right shows a large tank of water containing two regions of different depths.

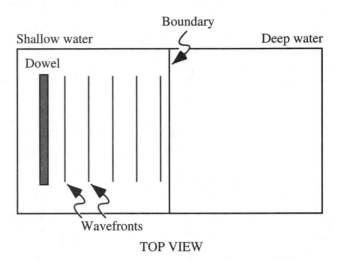

A periodic wave is being generated at the left side of the tank. At the instant shown, the wave has not yet reached the deeper water. (The lines in the diagram, called *wavefronts*, represent the *crests* of the wave.)

It is observed that water waves travel *more quickly* in deep water than in shallow water.

Make the following predictions based on what you have learned about the behavior of pulses passing from one spring to another.

1. Predict how the wavelengths of the waves in the two regions will compare. Explain.

2. Will a crest be transmitted as *a crest, a trough,* or *something in between?* Explain.

3. Predict how the frequencies of the waves in the two regions will compare. Explain.

⇨ Check your predictions with a tutorial instructor.

B. Suppose that the dowel were oriented as shown and rocked back and forth at a steady rate. (Only part of the tank is shown.)

On the diagram, (1) sketch the location and orientation of several wavefronts generated by the dowel, and (2) draw an arrow to show the direction of propagation of the wavefronts.

Ask a tutorial instructor for equipment that you can use to check your answer experimentally. (Generate a periodic wave by gently rocking the dowel back and forth at a steady rate.) If your answer was incorrect, resolve the inconsistency.

On the basis of your observations, how is the orientation of a straight wavefront related to its direction of propagation?

Explain how your answer can apply also to circular wavefronts (such as those made by a drop of water falling into a tank of water). Make a sketch of circular wavefronts to justify your answer.

Tutorials in Introductory Physics
McDermott, Shaffer, & P.E.G., U.Wash.

©Prentice Hall
Preliminary Edition, 1998

It is useful to represent straight wavefronts by drawing a single line along the direction that the wave moves. An arrowhead on the line (———➤———) indicates the direction of propagation. The line and arrowhead together are called a *ray,* and a diagram in which waves are represented by rays is called a *ray diagram.*

On the diagram at the bottom of the preceding page, draw a ray that shows the direction of propagation of the wave generated by the dowel.

C. Suppose that the dowel and the boundary between the shallow and deep water were oriented as shown.

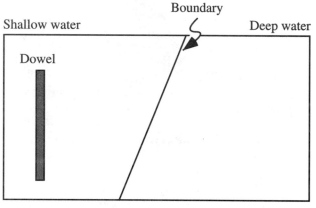

On the basis of your observations thus far, sketch two consecutive crests (1) before they cross the boundary, (2) as they are crossing the boundary, and (3) after they have crossed the boundary. (Ignore reflections at the boundary.)

Explain the reasoning you used in making your sketches.

D. Obtain a photograph that shows wavefronts incident from the left on a boundary between two regions of water and check your answers in part C.

1. Explain how you can tell from the photograph that the region of shallower depth is on the left-hand side of the photograph.

2. Describe how the wavefronts change in crossing the boundary. Use your answer to part B to determine how, if at all, the direction of propagation changes.

3. How does the phase of the incident wave change, if at all, in passing from one region to the other? (In other words, is a crest transmitted as *a crest, a trough,* or *something in between?*) Explain how you can tell from the photograph.

4. Are your predictions in part C consistent with your answers to the above questions? If not, resolve any inconsistencies.

II.   A water wave passing from deep water to shallow water

A. The diagram at right shows a periodic wave
    incident on a boundary between deep and
    shallow water. Assume that the wave speed
    in the shallow water is *half* as great as in the
    deep water.

Deep water

Shallow water

Ask a tutorial instructor for an enlargement
of the diagram and several transparencies.

1. Choose the transparency in which the parallel lines best represent the transmitted
    wavefronts.

    Explain the reasoning that you used to determine which set of parallel lines best
    represents the transmitted wave.

2. Place the transparency that you chose on the enlargement so that the parallel lines show
    the orientation and locations of the transmitted wavefronts.

    What criteria did you use to determine how to orient the transmitted wavefronts?

    Is there more than one possible orientation for the transmitted wavefronts that is
    consistent with your criteria?

3. Describe how the diagram would differ if the snapshot had been taken a quarter period
    later. (*Hint:* What is the direction of propagation of the transmitted wavefronts? How
    far do they travel in a quarter period?)

B. Sketch two diagrams below that illustrate waves passing from deep to shallow water at the
    angle of incidence shown. In one diagram, show the wavefronts; in the other, the rays.

Deep water

Shallow water

*Wavefront
diagram*

Deep

Shallow

*Ray
diagram*

The change that occurs in a wave when it propagates into a region with a different wave speed is called *refraction*. When representing waves by a ray diagram, the *angle of incidence* is defined as the angle between the ray that represents the incident wave and the *normal* to the boundary. The *angle of refraction* is defined analogously.

C. On the ray diagram in part B, label the angle of incidence, θ_i, and the angle of refraction, θ_r.

D. Obtain the equipment pictured at right from a tutorial instructor. The parallel lines on the sheet of paper represent the incident wavefronts (in deep water); the edge of the piece of cardboard represents the boundary between deep and shallow water. By changing the orientation of the sheet of paper, you can represent different angles of incidence.

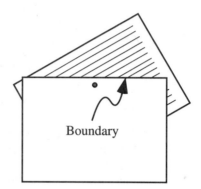

Boundary

Use the equipment to represent an angle of incidence of 0° and to show the refracted wavefronts that result.

When the angle of incidence is 0°, what is the angle of refraction?

As you gradually increase the angle of incidence, does the angle of refraction *increase, decrease,* or *stay the same?*

III. Summary

A. Each of the diagrams at right shows a ray incident on a boundary between two media.

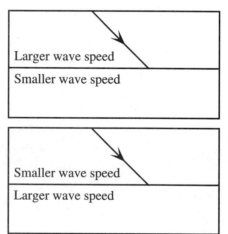

Larger wave speed

Smaller wave speed

Smaller wave speed

Larger wave speed

Continue each of the rays into the second medium. Using a dashed line, also draw the path that the wave would have taken had it continued without bending.

Does the ray representing a wave "bend" toward or away from the normal when:

• the wave speed is smaller in the second medium?

• the wave speed is larger in the second medium?

B. Does the ray representing a wave always "bend" when a wave passes from one medium into a different medium? If not, give an example when it does not "bend."

Tutorials in Introductory Physics
McDermott, Shaffer, & P.E.G., U.Wash.

©Prentice Hall
Preliminary Edition, 1998

ELECTROMAGNETIC WAVES

I.   Representations of electromagnetic waves

A. Shown below are pictorial and mathematical representations of an electromagnetic plane
wave propagating through empty space. The electric field is parallel to the z-axis; the
magnetic field is parallel to the y-axis.

$$\vec{E}(x, y, z, t) = E_o \sin(kx + \omega t)\,\hat{z} \qquad \vec{B}(x, y, z, t) = B_o \sin(kx + \omega t)\,\hat{y}$$

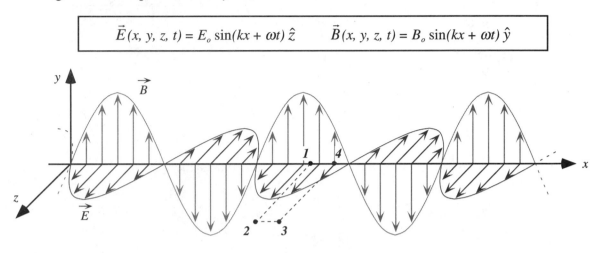

1. In which direction is the wave propagating? Explain how you can tell from the
expressions for the electric field and magnetic field.

Is the wave transverse or longitudinal? Explain in terms of the quantity or quantities that
are oscillating.

2. The points *1–4* in the diagram above lie in the *x–z* plane.

For the instant shown, rank the magnitude of the *electric field* at these points from largest
to smallest. If the electric field is zero at any point, state that explicitly.

Is your ranking consistent with the mathematical expression for the electric field
shown above? If not, resolve any inconsistencies. (For example, how, if at all, does
changing the value of z affect the value of $\vec{E}(x, y, z, t)$?)

For the instant shown, rank the magnitude of the *magnetic field* at points *1–4* from largest
to smallest. Check that your ranking is consistent with the expression for the magnetic
field, $\vec{B}(x, y, z, t)$, above.

3. In the diagram at right, the four points labeled "×" are all located in a plane parallel to the *y–z* plane. One of the labeled points is located on the *x*-axis.

    On the diagram, sketch vectors to show the direction and relative magnitude of the electric field at the labeled points.

    Justify the use of the term *plane wave* for this electromagnetic wave.

    ⇨ Check your answers to part A with a tutorial instructor.

B. Three light waves are represented at right. The diagrams are drawn to the same scale.

    1. How is the wave in case 1 different from the wave in case 2? Explain how you can tell from the diagrams.

    2. If the wave in case 2 were green light, could the wave in case 3 be *red* light or *blue* light? Explain.

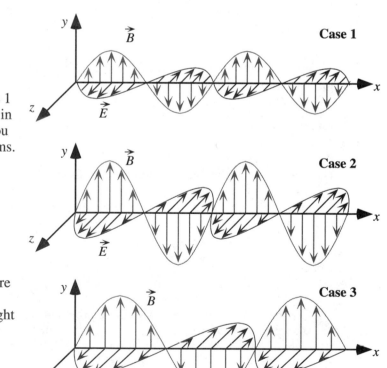

Tutorials in Introductory Physics
McDermott, Shaffer, & P.E.G., U.Wash.

©Prentice Hall
Preliminary Edition, 1998

II. Detecting electromagnetic waves

A. Write an expression for the force exerted on a charge, *q*, by (1) an electric field, $\vec{E}$, and (2) a magnetic field, $\vec{B}$.

If an electric field and a magnetic field were both present, would a force be exerted on the charge even if the charge were initially not moving? Explain.

B. A long, thin conducting wire (see figure at right) is placed in the path of a radio wave that can be represented as in section I.

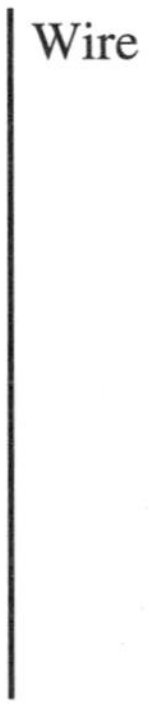

Wire

1. Suppose that the wire were oriented parallel to the *z*-axis.

   As the wave moves past the wire, would the *electric field* due to the radio wave cause the charges in the wire to move? If so, would the charges move in a direction along the length of the wire? Explain.

   As the wave moves past the wire, would the *magnetic field* due to the wave cause the charges in the wire to move in a direction along the length of the wire? Explain.

2. Imagine that the thin conducting wire is cut in half and that each half is connected to a different terminal of a light bulb. (See diagram at right.)

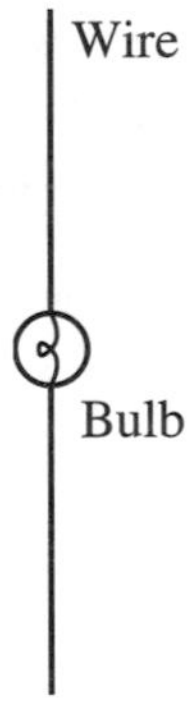

   Wire

   If the wire were placed in the path of the radio wave and oriented parallel to the *z*-axis, would the bulb ever glow? Explain. (*Hint:* Under what conditions can a bulb glow even if it is not part of a closed circuit?)

   Bulb

   How, if at all, would your answer change if the wire were instead oriented:

   • parallel to the *y*-axis? Explain.

   • parallel to the *x*-axis? Explain.

3. Suppose that the bulb were disconnected and that each half of the wire were connected in a circuit, as shown. (A conducting wire or rod used in this way is an example of an *antenna*.)

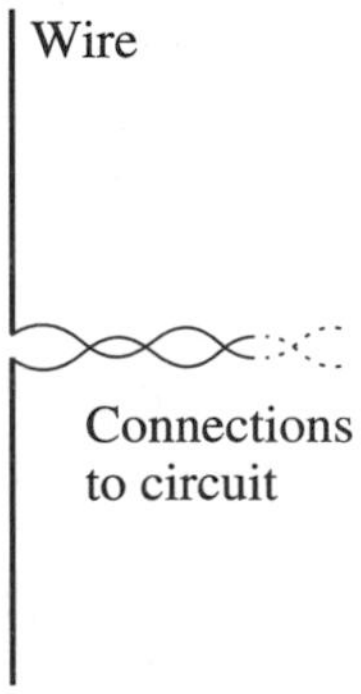

   Wire

   In order to best detect the oncoming radio wave (that is, to maximize the current through the circuit), how should the antenna be oriented relative to the wave? Explain.

   Connections to circuit

III. Electromagnetic waves and Maxwell's equations

A. Recall Faraday's law, $\oint \vec{E} \cdot d\vec{l} = -\dfrac{d\Phi_B}{dt}$, from electricity and magnetism. We shall consider

how each side of the equation for Faraday's Law applies to the imaginary loop,

$1 \rightarrow 2 \rightarrow 3 \rightarrow 4 \rightarrow 1$, in the figure for part A of section I.

1. For the instant shown in the figure, determine whether each quantity below is *positive, negative,* or *zero.* Explain your reasoning in each case.

   - the quantity $\int \vec{E} \cdot d\vec{l}$ evaluated over the path $1 \rightarrow 2$

   - the quantity $\int \vec{E} \cdot d\vec{l}$ evaluated over the path $2 \rightarrow 3$

   - the quantity $\oint \vec{E} \cdot d\vec{l}$ evaluated over the entire loop, $1 \rightarrow 2 \rightarrow 3 \rightarrow 4 \rightarrow 1$ (*Hint:* The answer is *not* zero!)

   For an imaginary surface that is bounded by a closed loop, it is customary to use the right-hand rule to determine the direction of the area vector that is normal to that surface. For example, the vector that is normal to the flat, imaginary rectangular surface bounded by the loop $1 \rightarrow 2 \rightarrow 3 \rightarrow 4 \rightarrow 1$ points in the *positive y*-direction.

2. At the instant shown in the figure, is the magnetic flux through the loop $1 \rightarrow 2 \rightarrow 3 \rightarrow 4 \rightarrow 1$ *positive, negative,* or *zero?* Explain how you can tell from the figure.

   A short time later, will the magnetic flux through the loop be *larger, smaller,* or *the same?* Explain how you can tell from the figure.

3. According to your answers in part 2 above, is the quantity $\left(-\dfrac{d\Phi_B}{dt} \right)$, written on the right-hand side of the equation for Faraday's law, *positive, negative,* or *zero?* Explain.

   According to your results in part 1 above, is the quantity on the left-hand side of this equation *positive, negative,* or *zero?*

   Do you get the same answer for both sides of the equation for Faraday's law? If not, resolve the inconsistencies.

B. Suppose that the electric field in a light wave were $\vec{E}(x, y, z, t) = E_o \sin(kx + \omega t)\, \hat{z}$.

   Would it be possible to have a magnetic field that is *zero* for all x and t? Use Faraday's law to support your answer. (*Hint:* How, if at all, would your answers in part A above be different if the magnetic field were zero for all x and t?)

Tutorials in Introductory Physics
McDermott, Shaffer, & P.E.G., U.Wash.

©Prentice Hall
Preliminary Edition, 1998

Geometrical optics

The activities in this tutorial should be performed in a darkened room. *In each experiment, make a prediction* before *you make any observations.* If you find that your predictions are incorrect, try to find the error in your explanation before continuing.

I.   Light

A.   Arrange a very small bulb, a cardboard mask, and a screen as shown at right. Select the largest circular hole (~1 cm in diameter) provided by the mask.

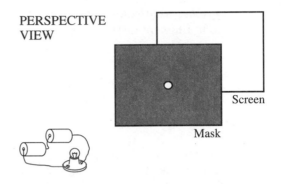

Predict what you would see on the screen. Explain in words and with a sketch.

Predict how moving the bulb upward would affect what you see on the screen. Explain.

Perform the experiments and check your predictions. If any of your predictions were incorrect, resolve the inconsistency.

B.   *Predict* how each of the following changes would affect what you see on the screen. Explain your reasoning and include sketches that support your predictions.

•   The mask is replaced by a mask with a *triangular* hole.

•   The bulb is moved farther from the mask.

Perform the experiments and check your predictions. Resolve any inconsistencies.

C.   *Predict* how placing a second bulb above the first would affect what you see on the screen. Explain.

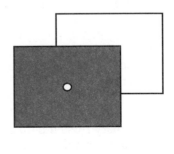
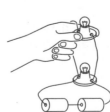
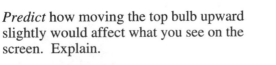

Predict how moving the top bulb upward slightly would affect what you see on the screen. Explain.

Perform the experiments. Resolve any inconsistencies.

D.   What do your observations suggest about the path taken by light from the bulb to the screen?

Tutorials in Introductory Physics
McDermott, Shaffer, & P.E.G., U.Wash.

©Prentice Hall
Preliminary Edition, 1998

E.   Imagine that you held a string of closely-spaced small bulbs one above the other. What would you expect to see on the screen?

   Predict what you would see on the screen if you used a bulb with a long filament instead. Explain.

Long filament bulb

   Check your prediction.

F.   *Predict* what you would see on the screen in the situation shown at right. Sketch your prediction below.

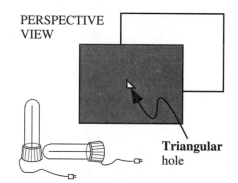

PERSPECTIVE VIEW

Triangular hole

   Compare your prediction with those of your partners. After you and your partners have come to an agreement, check your prediction. Resolve any inconsistencies.

G.   *Predict* what you would see on the screen in the situation pictured at right.

PERSPECTIVE VIEW

Triangular hole

   Predict what you would see on the screen if the mask were *removed*.

   Check your predictions. If any of your predictions were incorrect, resolve the inconsistency.

H.   *Predict* what you would see on the screen when an ordinary frosted bulb is held in front of a mask with a triangular hole as pictured at right.

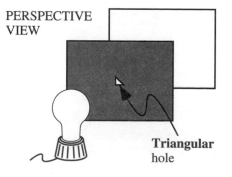

PERSPECTIVE VIEW

Triangular hole

   Obtain a frosted bulb from a tutorial instructor and check your prediction.

II.   Light: quantitative predictions

A. *Predict* the size of the lit region on the screen at right. Treat the bulb as a point source of light, *i.e.,* as if all the light emanates from a single point.

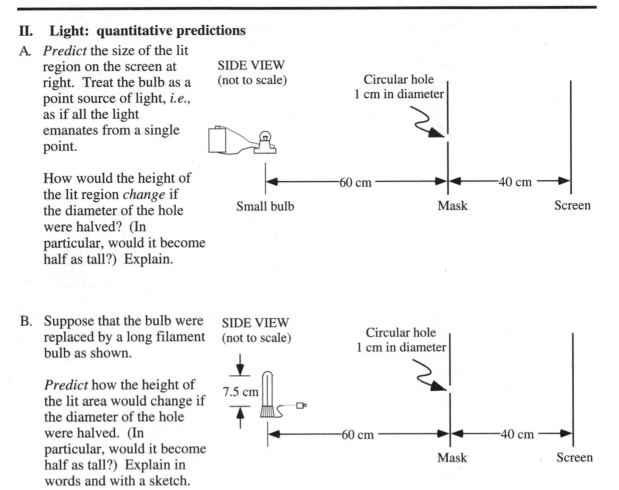

How would the height of the lit region *change* if the diameter of the hole were halved? (In particular, would it become half as tall?) Explain.

B. Suppose that the bulb were replaced by a long filament bulb as shown.

Predict how the height of the lit area would change if the diameter of the hole were halved. (In particular, would it become half as tall?) Explain in words and with a sketch.

Check your prediction. If your prediction was incorrect, resolve the inconsistency.

Predict the *approximate* height and shape of the lit region on the screen in the limit as the hole becomes very small, *e.g.,* the size of a pinhole. *(Hint:* In this limit, would the lit region be taller than, shorter than, or the same height as the *filament?)*

C. *Predict* what you would see on the screen in the situation pictured at right.

How would the height and width of each lit region *change* if the diameter of the hole in the mask were halved? Explain.

Check your predictions. If any of your predictions were incorrect, resolve the inconsistency.

III. Supplement: Shadows

Obtain a box, thread, and small bead (~5 mm in
diameter). Hang the bead as shown.

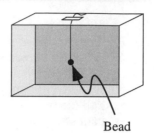

A. *Predict* what you would see on the screen at the
back of the box in the situation pictured at right.
Explain your reasoning.

Bead

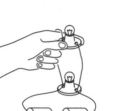

Predict how placing a second bulb above the
first would affect what you see on the
screen. Explain your reasoning.

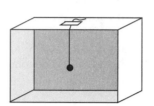

Perform the experiments. Resolve any
discrepancies between your observations
and predictions.

Must a region be *completely* without light for a shadow to be formed? Explain.

B. Suppose that the light from the top bulb in the situation above were red and the light from the
lower bulb were green. Predict what you would see on the screen. Explain.

Obtain red and green filters from a tutorial instructor and perform the experiment described
above. If your predictions were incorrect, find the error in your reasoning.

C. *Predict* what you would see on the screen in the
situation shown at right. Explain your reasoning.

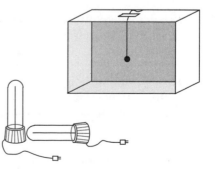

Suppose that the light from the vertical bulb were red
and the light from the horizontal bulb were green.
Predict what you would see on the screen.

Perform the experiments described above. Resolve any
discrepancies between your observations and
your predictions.

I.   The method of parallax

A. Close one eye and lean down in your chair so that your open eye is at table level. Have your partner drop a very small piece of paper (about 2 mm square) onto the table.

Hold one finger above the table and then move your finger until you think it is directly above the piece of paper. Move your finger straight down to the table and check whether your finger actually is in fact directly above the paper.

Try this exercise several times, with your partner dropping the piece of paper at different locations. Keep your open eye at table level. After several tries, exchange roles with your partner.

How can you account for the fact that when your finger misses the piece of paper, your finger is always either in front of the paper or behind it, but never to the left or right of the paper?

B. Suppose that you placed your finger behind the paper (as shown at right) while trying to locate the piece of paper.

Predict whether your finger would appear to be located *to the left of,* *to the right of,* or *in line with* the piece of paper if:

• you moved your head to the left.

• you moved your head to the right.

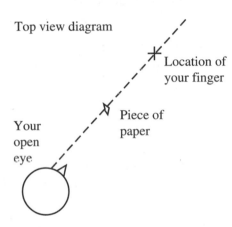

Top view diagram

Location of your finger

Piece of paper

Your open eye

Check your predictions. Resolve any inconsistencies.

C. Suppose that you had placed your finger in front of the piece of paper rather than behind it.

Predict whether the paper or your finger would appear on the left when you move your head to the left. Check your answer experimentally.

D. Devise a method based on your results from parts B and C by which you could locate the piece of paper. Your method should include how to tell whether your finger is directly over the piece of paper and, if not, whether it is in front of or behind the piece of paper. Describe your method to your partner, then test your method.

⇨ Check your method with a tutorial instructor.

Tutorials in Introductory Physics
McDermott, Shaffer, & P.E.G., U.Wash.

©Prentice Hall
Preliminary Edition, 1998

We will refer to the method that you devised for locating the piece of paper as the *method of parallax.*

II.   Image location

Obtain a small mirror and two identical nails. Place the mirror in the middle of a sheet of paper. Stand one nail on its head about 10 cm from the front of the mirror. We will call this nail the *object nail.*

Mirror

On the paper, mark the locations of the mirror and object nail.

A. Place your head so that you can see the image of the nail in the mirror.

Object nail

Use the method of parallax to position the second nail so that it is located in the same place as the image of the object nail. Mark this location on the paper.

Is the image of the nail located *on the surface of, in front of,* or *behind* the mirror? Explain.

Would observers at other locations agree that the image is located at the place you marked? Check your answer experimentally.

B. Move the nail off to the right side of the mirror as shown. Find the new image location.

Mirror

Object nail

In the following experiments, we will determine the location of an object and its image by a different technique called *ray tracing.* This technique is based on a model for the behavior of light in which we envision light being either emitted in all directions by a luminous object (such as a light bulb) or reflected in all directions by a non-luminous object (such as a nail).

III. Ray tracing

A. Place a large sheet of paper on the table. Stand a nail vertically at one end of the piece of paper.

Place your eye at table level at the other end of the piece of paper and look at the nail. Use a straightedge to draw a *line of sight* to the nail, that is, a line from your eye to the nail.

Repeat this procedure to mark lines of sight from three other very different vantage points, *then remove the nail.*

How can you use these lines of sight to determine where the nail was located?

What is the smallest number of lines of sight needed to determine the location of the nail?

B. Turn the large sheet of paper over (or obtain a fresh sheet of paper). Place the mirror in the middle of the sheet of paper, and place a nail in front of the mirror. On the paper, mark the locations of the mirror and the nail.

On the paper, draw several lines of sight to the *image* of the nail.

How can you use these lines of sight to determine the location of the image of the nail?

Use the method of parallax to determine the location of the image of the nail.

Do these two methods yield the same location of the image (to within reasonable uncertainty)?

C. Remove the mirror and the object nail. For each eye location that you used in part B, draw the path that light takes from the object nail to the mirror.

Draw an arrow head on each line segment (——————➤——————) to indicate the direction that light moves along that part of the path.

On the basis of the paths that you have drawn, formulate a rule that you can use to predict the orientation of the path that light takes after it is reflected by a mirror.

Tutorials in Introductory Physics
McDermott, Shaffer, & P.E.G., U.Wash.

©Prentice Hall
Preliminary Edition, 1998

D. Place the second nail at the location of the image of the object nail. Draw a diagram illustrating the path of the light from that nail to your eye for the same eye locations as above.

How is the diagram for this situation similar to the diagram that you drew in part C?

Is there any way that your eye can distinguish between these two situations?

IV. An application of ray tracing

In this part of the tutorial, use a straightedge and a protractor to draw rays as accurately as possible.

A. On the diagram at right, draw one ray from the pin that is reflected by the mirror.

If you were to place your eye so that you were looking back along the reflected ray, what would you see?

From one ray *alone* do you have enough information to determine the location of the image? If not, what can you infer about the location of the image from only a single ray?

B. On the diagram above, draw a second ray from the pin that is reflected by the mirror and that would reach an observer at a different location.

What can you infer about the location of the image from this second ray *alone?*

How can you use the two rays that you have drawn to determine the location of the image?

Is there additional information about the image location that can be deduced from three or more rays?

C. Determine the image location using the method of ray tracing from section III. (If it is necessary to extend a ray to show from where light appears to come, use a *dashed line*.)

Does the light that reaches the observer actually come from the image location or does this light only *appear* to come from that point?

What is the smallest number of rays that you must draw in using ray tracing to determine the location of the image of an object?

How does the distance between the mirror and the image location compare to the distance between the mirror and the pin?

The diagram that you drew above to determine the image location is called a *ray diagram*. The point from which the reflected light appears to come (*i.e.*, the location of the pin that you saw when you looked in the mirror) is called the *image location*. An image is said to be *virtual* when the light that forms the image does not actually pass through the image location. An image is said to be *real* when the light that forms the image does pass through the image location.

When drawing ray diagrams, use a solid line *with an arrow head* (⟶) to represent a ray, that is, a path that light takes. Use a dashed line (- - - - - - -) to extend a ray to show from where light *appears* to come in order to distinguish such a line from an actual ray.

©Prentice Hall
Preliminary Edition, 1998

I.   Cylindrical mirrors

A. A pin is placed in front of a cylindrical mirror as shown in the top view diagram at right. Lines *A–E* represent some of the light rays from the pin that reach the mirror. Points *M* and *N* represent the locations of two observers.

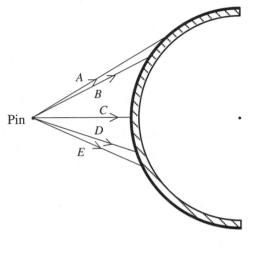

You have been provided with an enlargement of this top view diagram.

1. Use a ruler and a protractor to draw the reflected rays on the enlargement. (*Hint:* The center of the cylindrical mirror is marked on the diagram.)

Describe how you determined the direction of each reflected ray.

2. For each of the reflected rays, use a dashed line to show the direction from which the reflected ray appears to have come.

Do all of the reflected rays appear to have come from the same point?

3. On the diagram, draw a ray, *A'*, between rays *A* and *B*. Draw the corresponding reflected ray.

Which more nearly appear to pass through the same point: the reflected rays *A*, *A'*, and *B* or the reflected rays *A*, *B*, and *C?*

Tutorials in Introductory Physics
McDermott, Shaffer, & P.E.G., U.Wash.

©Prentice Hall
Preliminary Edition, 1998

Describe how, in principle, you could determine the location at which an observer at *M* would see an image of the pin. Label the approximate location on the diagram.

Determine and label the approximate location at which an observer at *N* would see an image of the pin.

Would the observers at *M* and *N* agree on the location of the image of the pin? Explain how you can tell from your ray diagram.

4. Ask a tutorial instructor for a semi-cylindrical mirror. Place the mirror on the enlargement and use the method of parallax to check your predictions. (You may find it helpful to tape the mirror onto the diagram.) If there are any inconsistencies between your predictions and your observations, resolve the inconsistencies.

B. Could you use any two rays (even those that do *not* pass near a particular observer) to find the location at which that observer sees the image of the pin in the case of:

• a plane mirror? Explain.

• a curved mirror? Explain.

Tutorials in Introductory Physics
McDermott, Shaffer, & P.E.G., U.Wash.

©Prentice Hall
Preliminary Edition, 1998

C. Observers at *M* and *N* are looking at an image of the
pin in the mirror.

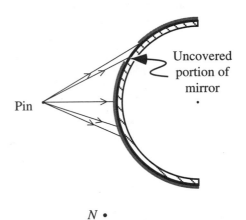

1. Suppose that all but a small portion of the
    mirror were covered as shown at right.

    How, if at all, would this change affect what the
    observers at *M* and *N* see? Explain.

    Determine the region in which an observer must
    be located in order to see an image of the pin.
    Discuss your reasoning with your partners.

    Would two observers at different locations in this region agree on the approximate
    location of the image? Explain.

2. Suppose that all but a small portion of the mirror
    near the center were covered, as shown at right.

    Determine the region in which an observer must be
    located in order to see an image of the pin.

    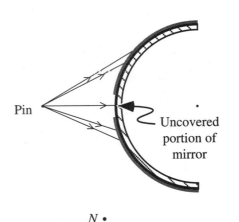

    Would two observers at different locations in
    this region agree on the approximate location of
    the image? If so, find the approximate image
    location. If not, explain how you can tell.

    Check your answers experimentally.

While the image location is independent of observer location in certain cases (*e.g.,* plane

mirrors), in general it is not. In many cases, however, it is possible to identify a limited range of

locations for which the image location is essentially independent of the observer location. An

example is when both the object and the observer lie very nearly along the axis of a cylindrical or

spherical mirror. In this situation, all rays are said to be *paraxial,* that is, they make small angles

with the axis of the mirror. Ray diagrams often specify the location of an image but not the

observer's location. For such a diagram, it should be assumed that the image location is

independent of the observer's location.

II.   Multiple plane mirrors

A. Stick a pin into a piece of cardboard and place two mirrors at right angles near the pin as shown in the top view diagram below.

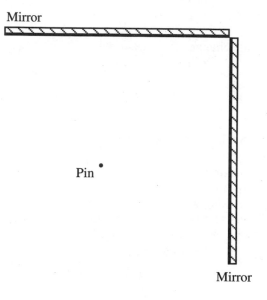

1. Describe what you observe.

2. View the arrangement from several locations and use the method of parallax to place a pin at each of the image locations.

3. Suppose that one of the mirrors were removed. Predict which image(s) you would still see and which image(s) would vanish.

Check your predictions. If any of your predictions were incorrect, resolve the conflict before continuing.

4. On the diagram above, sketch a ray diagram that accounts for each image.

Describe how one of the images differs from the others.

B. Gradually decrease the angle between the mirrors while keeping the pin between the mirrors.

How can you account for the presence of the additional images that you observe?

INTERPRETATION OF RAY DIAGRAMS

I.   Image location

A. A pin is held vertically at the back of a clear square container of water as shown at right. The portion of the pin below the surface of the water is not shown.

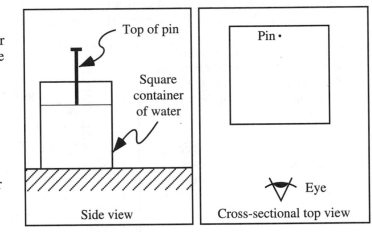

Side view

Cross-sectional top view

1. On an enlargement of the top view diagram, sketch several rays from the pin that pass through the water and out the front of the container, near the observer's eye.

For simplicity in answering the following questions, ignore the walls of the container (i.e., use the approximation that light passes directly from water to air, where it travels more quickly).

• On the basis of the rays that you have drawn, *predict* where the bottom of the pin would *appear* to be located to the observer. Discuss your reasoning with your partners.

• Would the bottom of the pin *appear* to be located closer to the observer, farther from the observer, or the same distance from the observer as the top of the pin? Explain.

2. Obtain the necessary equipment and use the method of parallax to check your predictions. If your ray diagram is not consistent with your observations, modify your ray diagram.

The place where the pin *appears* to be located is called the location of the image of the pin, or the *image location*.

3. In part 1, you assumed that light from the pin passes directly from water to air. Devise an experiment that would allow you to test whether this approximation is valid. (*Hint:* Use the method of parallax to see how the container *alone* affects the apparent location of the pin.)

Perform this experiment and check your answer.

Tutorials in Introductory Physics
McDermott, Shaffer, & P.E.G., U.Wash.

©Prentice Hall
Preliminary Edition, 1998

B. Suppose that a pencil were held vertically at the back of a circular beaker of water, as shown. (The portion of the pencil below the surface of the water is not shown.)

Is the image of the bottom of the pencil *closer to, farther from,* or *the same distance from* the observer as the top of the pencil?

Sketch a qualitatively correct ray diagram to support your answer. (*Note:* The center of the circle is marked by an "×.")

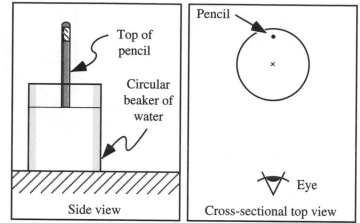

Use the method of parallax to place a second pencil at the location of the image of the bottom of the pencil, and check your predictions. If your prediction was incorrect, find your error.

C. Three students are discussing their results from part B:

Student 1: *"I think that the image is closer to me than the pencil itself. The closer something is, the bigger it looks. Because the image of the pencil appears wider than the pencil itself, the image must be closer to me than the pencil."*

Student 2: *"That sounds reasonable, but when I used parallax to determine the location of the image of the pencil, I found that it was farther from me than the pencil."*

Student 3: *"That doesn't make sense though. If the image were behind the pencil, then how could I see the image? Wouldn't the pencil block my view of the image?"*

Do you agree or disagree with each of these students? Discuss your reasoning with your partners.

Tutorials in Introductory Physics
McDermott, Shaffer, & P.E.G., U.Wash.

©Prentice Hall
Preliminary Edition, 1998

II.   Real and virtual images

Each of the ray diagrams below illustrates light from a pin that passes through a beaker of water. In one case, the pin is near the beaker of water; in the other case, far from the beaker.

A. For each case shown below, determine and label the location of the image of the pin. Explain how you determined your answer.

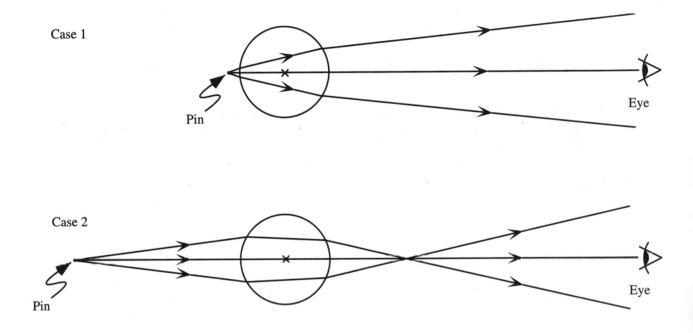

1. Use the ray diagrams above to answer the following questions:

In each case, which is farther from the observer:

- the image (below the water's surface) or
- the object (above the water's surface)?

In each case, which is farther from the observer:

- the image (below the water's surface) or
- the beaker of water?

2. Use the method of parallax to check your answers for both cases 1 and 2. Resolve any inconsistencies.

B. In each of the cases above, predict what you would see on a white paper screen placed at the image location. Imagine that the room has been darkened but that the pin is illuminated. (*Hint:* In either case, does the light from the pin that forms the image pass through the image location?)

Replace the pin with a lighted long-filament bulb and check your predictions. If either of your predictions were incorrect, resolve the inconsistency.

C. One of the images in part A is real; the other, virtual. Explain how you can tell which image is real and which is virtual on the basis of (1) the ray diagrams and (2) your observations in part B.

D. Explain how you can use a screen to determine the location of an image. In what cases, if any, would this method fail?

Tutorials in Introductory Physics
McDermott, Shaffer, & P.E.G., U.Wash.

©Prentice Hall
Preliminary Edition, 1998

In this tutorial, use a straightedge to draw lines that are meant to be straight.

I.   Convex lenses

A. Look at a very distant object through a convex lens. Hold the lens at arm's length so that you see a sharp image of the distant object.

Is the image formed by the lens *in front of, behind,* or *on the surface of* the lens? Use the method of parallax to determine the distance between the image and the lens.

B. Consider a point on the distant object that is also on the principal axis of the lens.

On the diagram below, sketch several rays from this distant point that reach the lens.

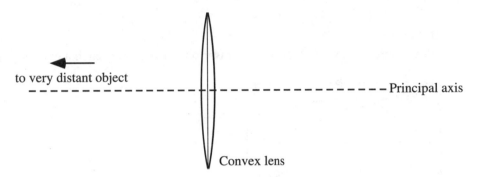

How are these rays oriented with respect to one another and to the principal axis? Explain.

On the basis of your observations from part A, show the continuation of each of these rays through the lens and out the other side. On the diagram, indicate where the rays converge.

Note: Refraction takes place at the two surfaces of the lens. However, in drawing a ray diagram for a thin lens, it is customary to treat rays as being refracted all at once at the center of the lens.

C. Suppose that you placed a very small bulb at the location of the image in part B.

How would the rays from the bulb that have passed through the lens be oriented? Draw a diagram to illustrate your answer. Explain.

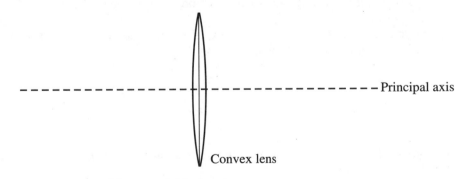

⇨   Discuss your answers with a tutorial instructor.

Tutorials in Introductory Physics
McDermott, Shaffer, & P.E.G., U.Wash.

©Prentice Hall
Preliminary Edition, 1998

The point of intersection of the principal axis and the image of a very distant object is called the *focal point*. The distance between the center of the lens and the focal point is called the *focal length*.

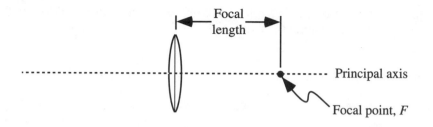

II. Ray tracing and convex lenses

The diagram below shows several rays from the eraser on a pencil that reach a converging lens.

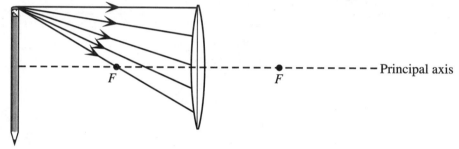

Pencil

A. Consider the ray that is parallel to the principal axis.

   Explain how you can use your observations from section I to draw the continuation of that ray on the right side of the lens. Draw this ray on the diagram.

B. Consider the ray that goes through the focal point on the left side of the lens.

   Explain how you can use your answers to part C of section I to draw the continuation of that ray on the right side of the lens. Draw this ray on the diagram.

C. How can you use these two rays to determine the location of the image of the eraser? On the diagram, label the image location.

Tutorials in Introductory Physics
McDermott, Shaffer, & P.E.G., U.Wash.

©Prentice Hall
Preliminary Edition, 1998

D. Consider the ray from the eraser that strikes the lens near its center, where the sides of the lens are nearly parallel.

Using the image location as a guide, draw the continuation of this ray through the lens and out the other side.

In your own words, describe the path of a ray that passes through the center of the lens.

E. Draw the continuation of the two remaining rays shown on the diagram through the lens and out the other side.

The rays that you drew in parts A, B, and D are called *principal rays,* and they are useful in determining the location of an image. In some cases, one or more of these rays may not actually pass through the lens; however, they may still be used in determining the image location. The principal rays are only a few of the infinitely many that we might draw from one point on the object.

F. On the diagram above, use the three principal rays from the *tip* of the pencil to determine the location of the image of the tip of the pencil. If possible, use a different color ink or pencil for this second set of rays.

G. The diagram below shows a small object placed near a convex lens. Draw all three principal rays and determine the location of the image. Clearly label the image location.

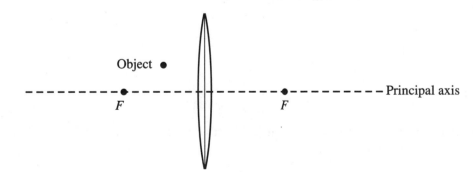

In your own words, describe how you knew to draw each ray.

III. Applications

A. A lens, a bulb, and a screen are arranged as shown below. A sharp, inverted image of the filament (not shown) appears on the screen when it is at the location shown.

Bulb   Lens   Screen

Predict how each of the following changes would affect what you see on the screen. Support your predictions with one or more ray diagrams.

- The screen is moved closer to or farther from the lens.

- The top half of the lens is covered by a mask.

    Does your answer depend on which side of the lens the mask is placed? If so, how? If not, why not?

B. Obtain the necessary equipment and check your predictions. In the space below, record how, if at all, your predictions were different from your observations. If your predictions were incorrect, resolve the inconsistencies.

C. If the screen were removed, would you still be able to see an image of the bulb? Does it matter where your eye is located?

    Turn off (but do not move) the bulb, remove the screen, and check your predictions. If your predictions were incorrect, resolve the inconsistencies. (*Hint:* Was a screen necessary to see an image in earlier situations, such as the situation in part A of section I?)

I.   Apparent size

A.   The diagram at right illustrates what an observer sees
when looking at two boxes on a large table.

From the diagram *alone:*

- is it possible to determine which box is closer to the observer?

- is it possible to determine which box *appears* wider to the observer?

- is it possible to determine which box *actually is* wider?

Discuss your reasoning with your partners.

B.   Obtain two soda cans and a cardboard tube that has a smaller diameter than the can.

1.   How can you arrange the two soda cans so that (a) they *appear* to be equally wide and
(b) one can *appears* wider than the other?

In the space below, draw a top view diagram for each case that can be used to compare
the apparent widths of the cans.

2.   How can you arrange one can and the tube so that (a) the two objects *appear* to be
equally wide and (b) the tube *appears* wider than the can?

In the space below, draw a top view diagram for each case that can be used to compare
the apparent widths of the two objects.

3.   What quantities affect the apparent size of an object? Describe how increasing or
decreasing each quantity affects the apparent size of that object.

Explain how you can use a top view diagram to determine whether one object *appears*
wider or narrower than another object to an observer at a particular location.

Tutorials in Introductory Physics
McDermott, Shaffer, & P.E.G., U.Wash.

©Prentice Hall
Preliminary Edition, 1998

II.   The image of an extended object

The ray diagram below shows a side view of a thin converging lens, a pencil, the image of the pencil, and five observer locations *(1–5)*. Two rays from the pencil tip are drawn through the lens.

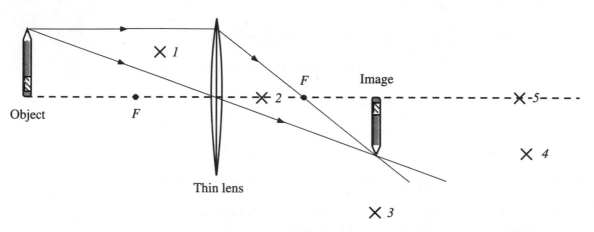

A.   Could an observer at each of the labeled points see a sharp image of the pencil tip (other than the actual pencil tip)? In each case, explain why or why not. Additionally, if an observer is able to see the image, indicate the direction that the observer would have to look to see the image.

- point *1*

- point *2*

- point *3*

- point *4*

- point *5*

B.   Use the above diagram to answer the following questions.

1.   From which of the labeled points could an observer see the image of the *eraser?* Draw rays to support your answer. (If possible, use a different color ink to draw these rays.)

From which point(s) could an observer see the entire image of the pencil? Explain.

To an observer at such a point, which would *appear* larger: the image of the pencil (with the lens in place) or the pencil (with the lens removed)? Explain how you can tell from the ray diagram.

Tutorials in Introductory Physics
McDermott, Shaffer, & P.E.G., U.Wash.

©Prentice Hall
Preliminary Edition, 1998

2. If you were to measure the length of the pencil and the length of the image using a ruler, which would actually be larger? Explain how you can tell from the ray diagram.

⇨ Check your results for section II with a tutorial instructor.

III. A magnifying glass

A. Obtain a convex lens. Use the lens as a *magnifying glass,* that is, to make an object such as a pencil appear larger. Start with the lens very close to the object.

Which is farther from you: the image or the object?

Where is the *object* relative to the lens and its focal points? (For example, is the object distance *greater than, less than,* or *equal to* the focal length of the lens?)

B. Draw a ray diagram that shows how to determine the location of the image that you observed above. Your diagram need not be drawn exactly to scale, but should correctly show the location of the object relative to the observer and to the lens and its focal points.

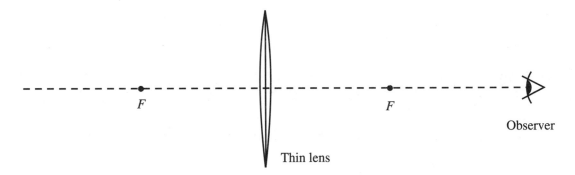

1. On the basis of your ray diagram, which is farther from the observer: the image or the object?

Is your answer consistent with your observations from part A? If not, resolve the inconsistency.

2. Does a magnifying glass simply make an object *appear* closer (*i.e.,* does it simply form an image of the object that is closer to you than the object itself)? If not, what does it do?

3. How can you tell from your ray diagram which would appear larger: the image of the pencil (with the lens in place) or the pencil (with the lens removed)?

IV. Magnification

A. The *lateral magnification, m,* is defined as $m = h'/h$, where h' and h represent the heights of the image and object, respectively. By convention, h' and h have opposite signs when the image is inverted.

Does the value of the lateral magnification depend on the location of the observer? Explain.

Consider the two examples in this tutorial. In each case, is the *absolute value* of the lateral magnification *greater than, less than,* or *equal to* one? (If your answer depends on observer location, choose an observer that can see the entire image.)

Does the lateral magnification tell you whether the image will *appear* larger than the object without the lens? Explain why or why not.

B. The *angular magnification, m_θ,* is defined as $m_\theta = \theta'/\theta$, where θ' and θ represent the angular sizes of the image and object, respectively. By convention, θ' and θ have opposite signs when the image is inverted.

Does the value of the angular magnification depend on the location of the observer? Explain.

Consider the two examples in this tutorial. In each case, is the *absolute value* of the angular magnification *greater than, less than,* or *equal to* one? (If your answer depends on observer location, choose an observer that can see the entire image.)

Does the angular magnification tell you whether the image will *appear* larger than the object without the lens? Explain why or why not.

⇨ Check your answers for section IV with a tutorial instructor.

Tutorials in Introductory Physics
McDermott, Shaffer, & P.E.G., U.Wash.

©Prentice Hall
Preliminary Edition, 1998

Physical optics

Tutorials in Introductory Physics
McDermott, Shaffer, & P.E.G., U.Wash.

I.   Periodic circular waves: single source

The circles at right represent wavefronts of a periodic circular
wave in a portion of a ripple tank. The dark circles represent
crests; the dashed circles, troughs. The diagram shows the
locations of the wavefronts at one instant in time, as a
photograph would.

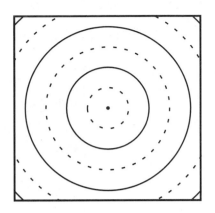

How, if at all, would the diagram differ:

• one-quarter period later? Explain.

• one period later? Explain.

II.   Periodic circular waves: two sources

A. The diagram at right
illustrates the wavefronts
due to each of two small
sources.

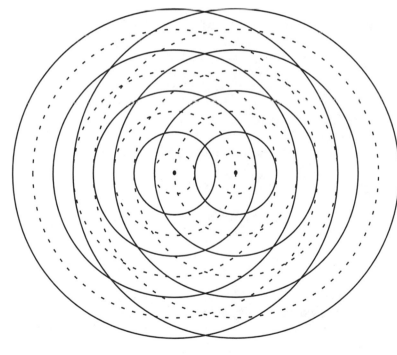

How do the frequencies
of the two sources
compare? Explain how
you can tell from the
diagram.

Are the two sources in
phase or out of phase
with respect to each
other? Explain how you
can tell from the
diagram.

What is the source separation? Express your answer in terms of the wavelength.

Tutorials in Introductory Physics
McDermott, Shaffer, & P.E.G., U.Wash.

©Prentice Hall
Preliminary Edition, 1998

B. Describe what happens at a point on the surface of the water where:

- a crest meets a crest

- a trough meets a trough

- a crest meets a trough

For each of the above cases, describe how your answer would differ if the amplitudes of the two waves were *not* equal. Explain your reasoning.

If the waves from two identical sources travel different distances to reach a particular point, the amplitudes of the waves from the two sources will not be the same at that point. For points that are sufficiently far from the sources, however, this difference in the amplitudes of the waves is small. For the remainder of this tutorial, we will *ignore* any such variations in amplitude.

C. You have been provided a larger version of the diagram of the wavefronts due to two sources.

Use different symbols (or different colors) to mark the places at which *for the instant shown:*

- the displacement of the water surface is zero (*i.e.,* at its equilibrium level)
- the displacement of the water surface is the greatest above equilibrium
- the displacement of the water surface is the greatest below equilibrium

(*Hint:* Look for patterns that will help you identify these points.)

What patterns do you notice? Sketch the patterns on the diagram in part A.

D. The representation that we have been using indicates the shape of the water surface at one particular instant in time.

Consider a point on your diagram where a crest meets a crest.

How would the displacement of the water surface at this point change over time? (*e.g.,* What would the displacement be one-quarter period later? What would it be one-half period later?)

Consider what happens at a point on your diagram where a crest meets a trough.

How will the displacement of the water surface at this point change over time?

E. Suppose that a small piece of paper were floating on the surface of the water. On the basis of your diagram, predict where the paper would move (1) the least, and (2) the most.

F. Consider a point where the water surface remains undisturbed.

1. Explain why that point *cannot* be the same distance from the two sources that we are considering.

For the two sources that we are considering, by how much must the distances from that point to the two sources differ?

Is there more than one possible value for the difference in distances? If so, list the other possible value(s) for the difference in distances. Explain.

2. Choose a variety of points where the water surface remains undisturbed.

For each of these points, determine the difference in distances from the point to the two sources. We will call this difference in distances ΔD. (This difference in distances is often called the *path length difference*.)

Divide all of the points where the water surface remains undisturbed into groups that have the same value of ΔD. Label each group with the appropriate value of ΔD, in terms of the wavelength, λ.

Justify the term *nodal lines* for groups of points that are far from the sources.

3. Similarly, group the points where there is maximum constructive interference according to their value of ΔD. We will call these *lines of maximum constructive interference*.

Label each group with the appropriate value of ΔD, in terms of the wavelength, λ.

4. Label each of the nodal lines and lines of maximum constructive interference with the corresponding value of $\Delta \varphi$, the phase difference between the waves from the two sources.

G. Imagine observing the waves from above the ripple tank. How, if at all, would the nodal lines and lines of maximum constructive interference change over time? Explain.

What patterns and symmetries do you notice in the arrangement of the nodal lines and the lines of maximum constructive interference?

H. Each of the photographs at right shows a *part* of a ripple tank that contains two sources that are in phase.

For each of the photographs, identify:

• nodal lines
• the approximate locations of the sources
• the line that contains the two sources

Which of the two photographs more closely corresponds to the situation that you have been studying? Explain your reasoning.

What difference(s) in the two situations could account for the difference in number and locations of the nodal lines?

I. Obtain a piece of paper and a transparency with concentric circles on them. The circles represent wavefronts generated by each of two point sources.

Suppose that the two sources are in phase and at the same location. Overlay the transparency on the paper to model this situation.

Explain why there are *no* nodal lines in this case.

Gradually increase the source separation until you first see nodal lines.

In the space at right, sketch the nodal lines and the lines of maximum constructive interference for this situation.

What is the source separation when this occurs?

Why can there be no nodal lines for a smaller source separation? Explain. *(Hint: For a given source separation, what is the largest possible value of ΔD?)*

Continue to increase the source separation and investigate how the source separation affects the number of nodal lines and their locations.

⇨ Check your answers above with a tutorial instructor.

I. Water waves incident on a single slit

Top view diagram

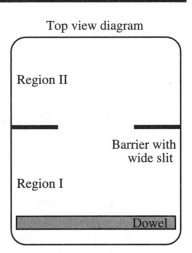

A. Obtain a pan of water and form a barrier in it that has a wide opening as shown. Place a dowel in the water and gently rock it back and forth to generate straight wavefronts at a constant rate.

Gradually decrease the width of the slit until it is completely closed. Observe the wavefronts in Region II as you make this change.

1. Describe how the shape of the wavefronts in Region II is affected as the width of the slit is decreased.

2. Compare the spacing of the wavefronts in the two regions (I and II). Is the spacing of the wavefronts in Region II affected by changing the width of the slit?

    Explain how your observation of the spacing of the wavefronts is consistent with the relative wave speeds in the two regions.

3. How, if at all, does the amplitude of the wave in Region II change when the slit is made slightly narrower? In particular, consider two cases in which:

    • the slit is initially very wide and is made slightly narrower.

    • the slit is initially very narrow and is made even narrower.

B. It is difficult to make periodic waves using the equipment at your table. Ask a tutorial instructor for photographs of periodic waves incident on slits of various widths.

1. Are the wavefronts in the photographs consistent with your observations above?

2. Identify the picture(s) in which the slit acts most like a point source of water waves. Explain.

    How could you modify the situation in this photograph in order to make the slit act more like a point source of waves?

3. Identify the photograph(s) in which the slit does not significantly affect the shape of the wavefronts.

    How could you modify the situation in this photograph so that the slit affects the wavefronts that pass through the slit to an even lesser extent?

Tutorials in Introductory Physics
McDermott, Shaffer, & P.E.G., U.Wash.

©Prentice Hall
Preliminary Edition, 1998

As you have observed, the behavior of waves passing through a slit can depend on the size of the slit. For the remainder of this tutorial, we will consider the case of waves passing through two very narrow slits.

II.   Water waves incident on two very narrow slits

For this part of the tutorial, you will not be asked to perform any experiments.

A.   A periodic water wave is incident on a barrier with two identical narrow slits. Each slit is narrow enough so that it may be treated as a (single) source of circular wavefronts.

Describe the shape of the wavefronts that emanate from each slit.

Top view diagram (not to scale)

B.   Obtain an enlargement of the diagram at right that shows the wavefronts for the case in which the distance between the centers of the slits is 3λ.

For this situation, which values of ΔD (the difference in distances from a point to each of the slits) correspond to (1) nodal lines and (2) lines of maximum constructive interference? Explain.

At how many points along the line X–X' in the diagram above is there (1) complete destructive interference and (2) maximum constructive interference? Mark the *approximate* locations of all of these points on the diagram above, and label each point with the corresponding value of ΔD. Assume that the tank is very wide and that the line X–X' is very far from the slits. Explain.

C.   Suppose that the width of *one* of the slits were decreased (without changing the distance between the centers of the slits). How, if at all, would this modification affect how much the water surface would move at the points you marked above? Explain your reasoning. (*Hint:* How can you use your observations from part A of section I in this case?)

Thus far we have observed the behavior of water waves when they pass through narrow slits.

Below we investigate the behavior of light passing through two very narrow slits.

Tutorials in Introductory Physics
McDermott, Shaffer, & P.E.G., U.Wash.

©Prentice Hall
Preliminary Edition, 1998

III. Light incident on two narrow slits

A. Red light from a distant point source is incident on a mask with two identical, narrow vertical slits. The photograph at right illustrates the pattern that appears at the center of a distant screen.

Pattern on screen

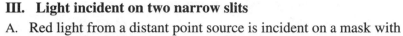

How does this pattern *differ* from what you would have predicted *if* you had used the idea that light travels in straight lines through slits?

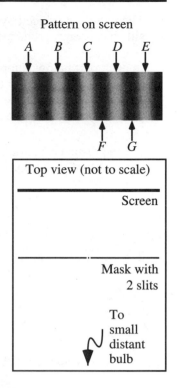

Top view (not to scale)

Screen

Mask with
2 slits

To
small
distant
bulb

B. Compare the situation in part II (in which a periodic water wave was incident on two identical, narrow slits) to the experiment described above.

Which points along line *X–X'* in the ripple tank best correspond to:

- points of minimum intensity *(e.g., points F and G)?* Explain.

- points of maximum intensity *(e.g., points A–E)?* Explain.

For a point of minimum intensity *(e.g., points F and G),* identify the quantity or quantities that are adding to zero at that point. Explain your reasoning.

C. For each of the lettered points, determine ΔD (in terms of λ) and $\Delta\varphi$, the phase difference between the waves. Record your answers below. *Note:* Point *C* is at the center of the screen.

	point *A*	point *B*	point *C*	point *D*	point *E*	point *F*	point *G*
ΔD							
$\Delta\varphi$							

D. Suppose that one of the slits were covered.

At which, if any, of the points *A–G* would the brightness increase? Explain.

At which, if any, of the points *A–G* would the brightness decrease? Explain.

In the space at right, sketch the pattern that would appear on the portion of screen shown in the above photograph when one of the slits is covered. Explain.

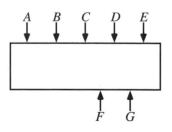

E. The pattern produced by red light passing through two very narrow slits has been reproduced at right.

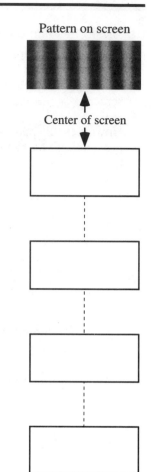

Pattern on screen

Center of screen

In each part below, suppose that a *single* change were made to the original apparatus. For each case, determine how, if at all, that change would affect the pattern on the screen. Sketch your predictions in the spaces provided.

1. the distance between the slits is decreased (without changing the width of the slits)

2. the screen is moved closer to the mask containing the slits

3. the wavelength of the incident light is decreased

4. the width of each slit is decreased (without changing the distance between the slits)

F. Consider point *B*, the first maximum to the left of the center of the screen.

Pattern on screen

Suppose that the two slits are separated by 0.2 mm, that the screen is 1.2 m away from the slits, and that the distance from the center of the pattern (point *C*) to point *B* is 3.6 mm.

Use this information to determine the wavelength of the light. Describe any approximations that you make in answering this question.

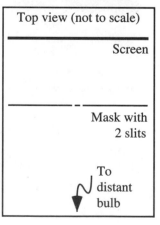

Top view (not to scale)

Screen

Mask with 2 slits

To distant bulb

Tutorials in Introductory Physics
McDermott, Shaffer, & P.E.G., U.Wash.

©Prentice Hall
Preliminary Edition, 1998

I.   Two-slit interference

A.   Red light from a distant point source is incident on two very narrow identical slits, S_1 and S_2, separated by a distance d. The photograph at right illustrates the pattern that appears on a distant screen.

Two-slit pattern on screen

The magnified view shows the path from slit S_1 to point X, a point on a distant screen.

On the magnified view:

• Draw an arrow to show the direction from slit S_2 to point X.

• Identify and label the line segment of length ΔD that represents how much farther light travels from one slit than from the other to reach point X.

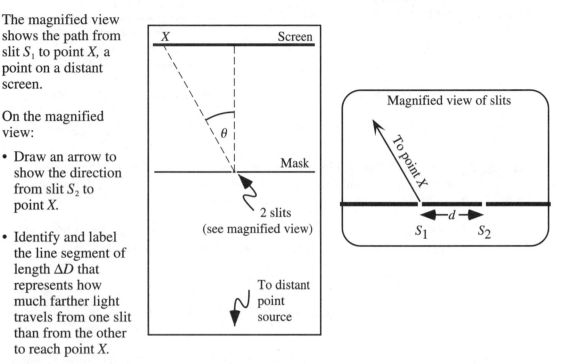

B.   In a previous homework, you found an expression for ΔD in terms of d and θ that was valid for points far from two point sources. Using that expression, write equations (in terms of λ, θ, and d) that you can use to calculate the angle(s) for which there will be:

• maximum constructive interference *(i.e., a maximum)*

• complete destructive interference *(i.e., a minimum)*

C.   Suppose that the screen were semicircular, as shown.

On the diagram, mark the locations of *all* minima and maxima for the specific case $d = 2.4\lambda$.

Label each maximum and minimum with the corresponding value of:

• ΔD,
• θ, and
• $\Delta \varphi$, the phase difference between the waves from the two slits.

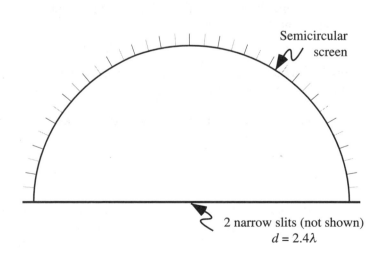

Semicircular screen

2 narrow slits (not shown)
$d = 2.4\lambda$

Tutorials in Introductory Physics
McDermott, Shaffer, & P.E.G., U.Wash.

©Prentice Hall
Preliminary Edition, 1998

II.   Three-slit interference

A. Consider a point M on the distant screen where there is a *maximum* due to the light from S_1 and S_2.

If a third slit were added as shown at right, would there *still* be maximum constructive interference at point M? Explain.

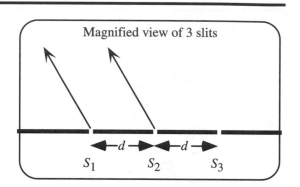

Magnified view of 3 slits

S_1 S_2 S_3
$\leftarrow$$d$$\rightarrow$ $\leftarrow$$d$$\rightarrow$

Suppose that more identical slits were added with all adjacent slits a distance d apart.

Would there still be maximum constructive interference at point M? Explain.

Let ΔD_{adj} represent the difference in distances from two adjacent slits to a location on the screen. For two slits, you found that any point of maximum constructive interference is farther from one of the slits by a whole number of wavelengths (*i.e.*, ΔD is 0, λ, 2λ, ...).

For three or more evenly-spaced slits, what is the corresponding condition for locations of maximum constructive interference? Express this condition in terms of ΔD_{adj}.

We will call a location at which light from *all* of the slits is in phase a *principal maximum*.

B. Consider a point N on the screen where there is a *minimum* due to the light from S_1 and S_2.

Will the screen remain completely dark at point N after the third slit is added as shown above? If not, will point N be as bright as a principal maximum? Explain.

C. Obtain a set of transparencies of sinusoidal curves. Each transparency can represent the light from a single narrow slit. In particular, what quantity or quantities can these curves be used to represent?

Find a way to align three sinusoidal curves so they would add in a way that results in a minimum for three slits.

What is the smallest value of ΔD_{adj} that corresponds to a minimum for three slits?

Would twice this value also correspond to a minimum? three times? four times?

Write out the first few values of ΔD_{adj} that correspond to minima for three slits. Write out enough values to clearly indicate the pattern.

How many minima are there between adjacent principal maxima for three slits?

D. On the diagram at right, mark the locations of *all* minima and principal maxima for the specific case of three identical slits separated by a distance $d = 2.4\lambda$.

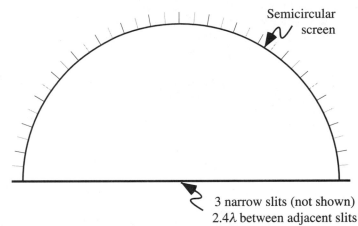

Semicircular
screen

3 narrow slits (not shown)
2.4λ between adjacent slits

Label each minimum and principal maximum with the corresponding value of:

- ΔD_{adj},
- θ, and
- $\Delta\varphi_{adj}$, the phase difference between the waves from *adjacent* slits.

Compare and contrast this sketch with your sketch from part C of section I for the case of light incident on two slits separated by $d = 2.4\lambda$. Discuss the similarities and differences.

III. Multiple-slit interference

A. Suppose that coherent red light were incident on a mask with four narrow slits a distance d apart.

Use the transparencies of sinusoidal curves to find the smallest value of ΔD_{adj} that would correspond to a minimum for this case.

Which integer multiples of this value of ΔD_{adj} would correspond to other minima? Which would not?

Which values of ΔD_{adj} would correspond to the principal maxima?

How many minima would there be between adjacent principal maxima?

B. Generalize your results from the 2-slit, 3-slit, and 4-slit cases to determine the smallest value of ΔD_{adj} that would correspond to a minimum for the case of N identical, evenly-spaced slits.

Which integer multiples of this value of ΔD_{adj} would correspond to other minima? Which would not?

How many minima would there be between adjacent principal maxima?

C. Coherent red light is incident on a mask with two very narrow slits a distance *d* apart. The photograph at right illustrates the pattern that appears on a distant screen.

On the photograph, label each of the maxima and minima with the corresponding value of ΔD_{adj}.

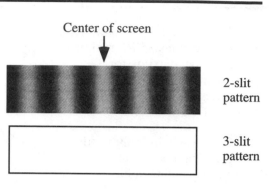

Center of screen

2-slit pattern

3-slit pattern

Suppose that a *third* slit were added to the mask so that adjacent slits were separated by the same distance *d* as before.

In the space provided, sketch the pattern that you would expect to see on the same part of the screen. On your sketch, clearly label each minimum and principal maximum with the corresponding value of ΔD_{adj}.

Ask a tutorial instructor for photographs that illustrate the patterns that appear on a distant screen when light is incident on two masks: one with two slits and one with three slits.

D. How would the pattern differ if the mask contained *four* slits separated by the same distance *d* as before? Sketch your prediction in the space provided at right. Explain your reasoning.

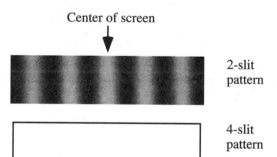

Center of screen

2-slit pattern

4-slit pattern

How would the pattern differ if the mask contained *five* slits separated by the same distance *d* as before? Explain your reasoning.

Ask a tutorial instructor for photographs that illustrate the patterns that appear on a distant screen when light is incident on masks with different numbers of slits.

I. Determining the location of the first minimum for many slits

A. Red light from a distant point source is incident on a mask with ten identical, evenly-spaced, very narrow slits. (See diagrams at right and below.)

On the magnified view below, label the line segment of length ΔD_{adj} that represents how much farther light must travel from slit 1 than from slit 2 to reach point X on a distant screen.

What is the *smallest* value of ΔD_{adj} that corresponds to a minimum for 10 slits? (Transparencies of sine curves are available in case you would like to review these concepts.)

The minimum that corresponds to this smallest value of ΔD_{adj} is called the *first minimum*.

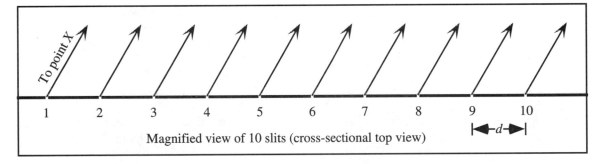

Magnified view of 10 slits (cross-sectional top view)

B. Suppose that point X marks the location of the first minimum on the screen.

How much farther (in terms of λ) does the light from slit 1 travel than the light from slit 3 in reaching point X? Explain.

C. Suppose that only slit 1 is uncovered, and all other slits 2–10 are covered. Which other slit could be uncovered so that the screen would be completely dark at point X? Explain.

Suppose that this *pair* of slits is uncovered, so that point X is completely dark. If slit 2 were now uncovered, would point X remain completely dark? If not, which other slit could also be uncovered (to pair with slit 2) so that point X once again becomes completely dark? Explain.

Tutorials in Introductory Physics
McDermott, Shaffer, & P.E.G., U.Wash.

©Prentice Hall
Preliminary Edition, 1998

D. Show how you could group all ten slits into five pairs of slits so that the light waves from each pair add to zero at point *X*.

E. Suppose that the number of slits is doubled and the distance between adjacent slits is halved. (See below.) The new slits are labeled *1a–10a*. (The diagram uses the same scale as the preceding one.)

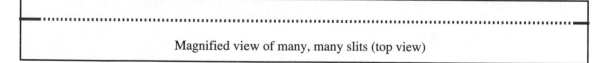

1   *1a*   2   *2a*   3   *3a*   4   *4a*   5   *5a*   6   *6a*   7   *7a*   8   *8a*   9   *9a*   10   *10a*

Magnified view of 20 slits (top view)

Would the first minimum in this case be located at the same angle θ as in part B? Explain.

F. If we continued to add slits in this way (*i.e.,* doubling the number of slits, but halving the distance between adjacent slits), would the angle to the first minimum change? Explain.

When the number of slits becomes very large as shown below, how can the slits be paired to determine the angle to the first minimum?

Magnified view of many, many slits (top view)

Tutorials in Introductory Physics
McDermott, Shaffer, & P.E.G., U.Wash.

©Prentice Hall
Preliminary Edition, 1998

II. Motivation for a model for single-slit diffraction

The photograph below illustrates the pattern that appears on a distant screen when light from a distant point source passes through a single narrow vertical slit. This pattern is an example of a *single-slit diffraction pattern*.

A. How is this pattern *different* from what you would predict using the ideas developed in geometrical optics (*e.g.,* light travels in straight lines through slits)?

The presence of minima in a diffraction pattern suggests that diffraction is an interference phenomenon. We can model single-slit diffraction as follows: Consider the slit as consisting of many identical, very narrow, evenly–spaced "slits" that are so close to one another that the edges of these "slits" meet. The interference pattern produced by the light passing through the many "slits" approximates the single-slit diffraction pattern.

B. Consider the following dialogue between two students:

Student 1:   *"I don't see why there are minima when there's only a single slit—don't you need two waves to have destructive interference?"*

Student 2:   *"You can model the single slit as many identical smaller interfering 'slits,' each small enough to act like a point source. The first minimum occurs where the light waves from the two 'slits' at the edges of the single slit are 180° out of phase."*

Do you agree with student 2's response to student 1? Discuss your reasoning with your partners.

III. Applications of the model

A. The photograph at right shows the diffraction pattern produced on a distant screen by green light incident on a narrow slit.

Label the point(s) that correspond to the first minimum with an *X*. Which points correspond to higher-order minima?

Suppose that red light, instead of green light, were incident on the same slit.

Determine whether the angle to the first minimum for red light would be *larger than, smaller than,* or *the same as* for green light. In the space below, draw diagrams that support your prediction, and explain your reasoning.

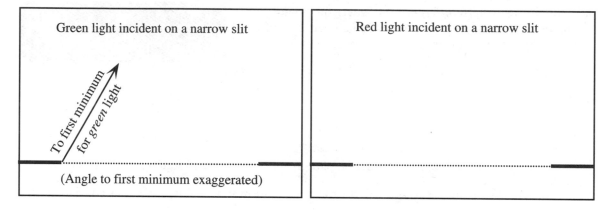

Would you expect the locations of the higher-order minima to change? If so, how?

In the space provided below the photograph above, show how the diffraction pattern would be different if red light, rather than green light, were incident on the narrow slit.

Ask a tutorial instructor for the color photograph that shows the diffraction patterns produced by red light and by green light on a narrow slit so that you may check your predictions.

B. The photograph at right shows the diffraction pattern produced by laser light incident on a narrow slit.

Narrow slit

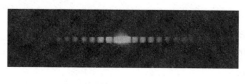

Use the model that we have developed to predict how the pattern would change if the slit were made *even narrower*.

Even narrower slit

In the space below the photograph, sketch your prediction clearly indicating how the pattern would change.

Ask a tutorial instructor for the photograph showing diffraction patterns produced by light incident on a narrow slit and on an even narrower slit so that you may check your predictions.

Tutorials in Introductory Physics
McDermott, Shaffer, & P.E.G., U.Wash.

©Prentice Hall
Preliminary Edition, 1998

C. Describe what you would see on the screen if the width of the slit were gradually decreased to zero. Discuss your predictions with your partners.

D. If a diffraction pattern has several minima (like the patterns shown in this tutorial), is the width of the slit *greater than, less than,* or *equal to* λ? Explain your reasoning.

E. In part A, you drew a diagram that showed how to find the angle to the first minimum for green light incident on a narrow slit. Use your diagram to determine whether the width of the slit was *greater than, less than,* or *equal to* the wavelength of the incident light in that case.

Is this comparison consistent with your answer to part D? If not, resolve the inconsistency.

F. Use the model that we have developed to write an equation that can be used to determine the angle to the first minimum in the case of single-slit diffraction.

Explain how you can account for the fact that the above equation, which describes the location of a minimum in the case of single-slit diffraction, is similar in appearance to the equation that describes the location of a maximum in the case of two-source interference.

©Prentice Hall
Preliminary Edition, 1998

I. Single-slit diffraction

Monochromatic light from a distant point source is incident on a mask that contains a single narrow vertical slit. The photograph at right shows the pattern produced on a distant semicircular screen. The corresponding graph of relative intensity is shown above the photograph, where *relative intensity* is intensity divided by the intensity at $\theta = 0$ (*i.e.,* $I(\theta)/I_{max}$).

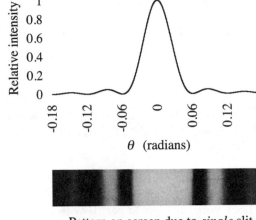

Pattern on screen due to *single* slit

A. The minima that occur in the case of a single slit are called *diffraction minima*. On the photograph and on the graph, identify the locations of the diffraction minima.

B. Consider the following dispute between two physics students:

Student 1: *"In lab, I determined that the width of one of the slits that we used to study single-slit diffraction was about 0.1 mm—that's definitely larger than λ."*

Student 2: *"You must have made a mistake. A diffraction pattern has minima only when the slit width is less than λ."*

Do you agree or disagree with each of these students? Explain your reasoning.

II. Combined interference and diffraction

A. A second slit, identical in size to the first, is cut in the mask. The distance between the centers of the slits, *d,* is equal to 50λ.

What would you see on the screen if the original slit were covered and the second slit were uncovered?

B. Imagine that both slits are now uncovered.

For what angles would the light from one slit be 180° out of phase with the light from the other slit? (*Hint:* For small angles, $\sin\theta \approx \theta$, where θ is in radians.)

On the relative intensity graph above, clearly label these angles.

Tutorials in Introductory Physics
McDermott, Shaffer, & P.E.G., U.Wash.

©Prentice Hall
Preliminary Edition, 1998

When the second slit is uncovered, does the intensity at the locations of the diffraction minima *increase, decrease,* or *stay the same?* Explain your reasoning.

When the second slit is uncovered, how does the pattern that you see on the screen change? In the space below the photograph on the first page, clearly show how the pattern would be different.

⇨ Check your answers to part B with a tutorial instructor.

C. Suppose that the width of both slits, *a*, were gradually decreased (while keeping the distance between the centers of the slits the same).

Which minima would move as *a* is decreased?

Choose two or more relative intensity graphs below that illustrate such a change. (Enlargements of these graphs have also been provided.)

D. Suppose instead that the distance between the centers of the slits, *d*, were gradually decreased (while keeping the widths of the slits the same).

Which minima would move as *d* is decreased?

Choose two or more relative intensity graphs below that illustrate such a change.

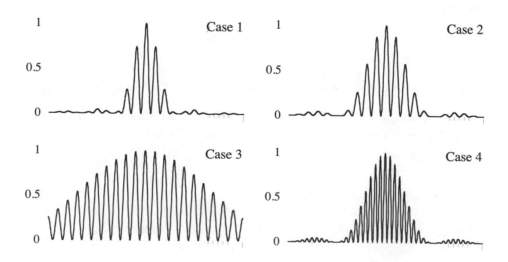

Tutorials in Introductory Physics
McDermott, Shaffer, & P.E.G., U.Wash.

©Prentice Hall
Preliminary Edition, 1998

The minima that occur when only one slit is open are called diffraction minima. The minima that occur where the light from one slit is 180° out of phase with the light from the other slit are called *interference minima*.

E. On each of the four graphs in part D:

- Clearly label (1) the interference minima that are closest to the center of the pattern and (2) the diffraction minima that are closest to the center of the pattern.

- Sketch the graph of relative intensity that would result if one of the slits were covered.

For each of the four cases, is your relative intensity graph consistent with the minima that you identified? If not, resolve any inconsistencies.

⇨ Check your answers to parts C–E with a tutorial instructor.

F. Consider the relative intensity graph shown at right.

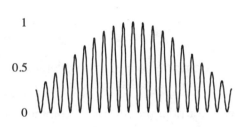

Suppose that both slits were made narrower (without changing the distance between the centers of the slits).

On the figure at right, show how the relative intensity graph would change. Explain your reasoning.

Suppose that after gradually narrowing both slits, one of the slits were then covered. In the space at right, sketch the relative intensity graph for this case.

How does your graph compare to what you would expect for a *point source?* If it is different, how could you modify the physical situation so that the relative intensity graph better approximates that due to a point source?

In order for the relative intensity graph to be a good approximation of that due to a point source, how must the width of the slit compare to λ? Explain your reasoning.

III. Quantitative predictions

Consider the following relative intensity graph for a double-slit experiment. The wavelength of the light used was $\lambda = 633$ nm.

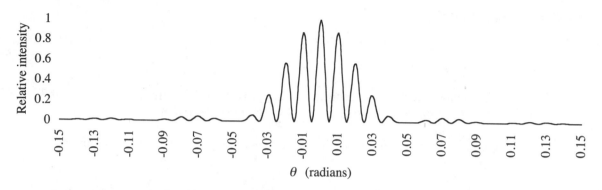

A. Determine the width of the slits and the distance between the slits. Clearly indicate which features from the graph you are using.

Compare your results with those obtained by your partners. If your answers are different, resolve any discrepancies.

B. Consider the following comment made by a student:

> *"To determine slit width, I used the first minimum, at $\theta = 0.005$ radians, and to determine the distance between the slits, I used the first maximum, at $\theta = 0.01$ radians."*

What is the flaw in the reasoning used by this student? Explain your reasoning.

C. You may have already noticed that the maxima are (approximately) 0.01 radians apart, except that there are no maxima at $\theta = 0.05$ radians or $\theta = 0.10$ radians.

How can you account for these "missing" maxima? (*Hint:* Consider how the relative intensity graph would be different if the width of the slits were decreased.)

Are your answers from part A consistent with your answer above? If not, resolve the inconsistency.

I. Transmission and reflection at a boundary

The sketches below show a pulse approaching a boundary between two springs. In one case, the pulse approaches the boundary from the left; in the other, from the right. The springs are the same in both cases, and the linear mass density is greater for the spring on the right than for the spring on the left.

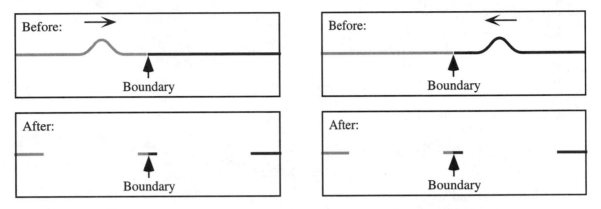

Complete the sketches to show the shape of the springs a short time after the trailing edge of the pulse shown has reached the boundary. Be sure to show correctly (1) the relative widths of the pulses and (2) which side of the spring each pulse is on. (Ignore relative amplitudes.)

> Compare your diagrams with those of your partners. Resolve any inconsistencies.

II. Thin-film interference

You may have observed that when a beam of light strikes a piece of glass, it is partially reflected and partially transmitted, similar to the behavior of a pulse on a spring when it reaches the junction between two connected springs.

In this tutorial, we will consider a beam of light in air ($n = 1$) incident on a soap film ($n = {}^4/_3$). We will make an analogy between this situation and a pulse incident on a boundary between two springs of different mass densities.

A. In this analogy, would the soap film better correspond to the spring with the larger linear mass density or the smaller linear mass density? Explain your reasoning.

> Discuss your reasoning with your partners.

B. When comparing two materials of different indices of refraction, the material with the higher index of refraction is sometimes said to be more "optically dense" than the other. Is this terminology consistent with the analogy that you made in part A?

Tutorials in Introductory Physics
McDermott, Shaffer, & P.E.G., U.Wash.

©Prentice Hall
Preliminary Edition, 1998

C. Consider light incident on a *thin* soap film, as illustrated in the cross-sectional side view diagram at right.

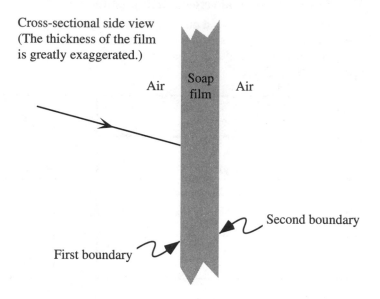

Cross-sectional side view (The thickness of the film is greatly exaggerated.)

The soap film is supported by a loop (not shown), which is held vertically. Only a small portion of the film has been shown; the thickness of the film is greatly exaggerated.

In answering the following questions, use the analogy between this situation and the connected springs.

1. Reflection and transmission at the first boundary

  a. On the diagram, draw rays that correspond to the light that is transmitted and reflected at the first boundary (on the left).

  b. Is the frequency of the transmitted wave (in the film) *greater than, less than,* or *equal to* the frequency of the incident wave (in the air)?

  c. Is the wavelength of the transmitted wave (in the film) *greater than, less than,* or *equal to* the wavelength of the incident wave (in the air)?

  d. For light incident on the first boundary, would the reflection at this boundary be more like reflection from a *fixed end* or from a *free end?* Explain.

  e. On the basis of your answers above:

   At the first boundary, would the reflected wave be *in phase* or *180° out of phase* with the incident wave (*i.e.,* is there a phase change upon reflection)?

   At the first boundary, would the transmitted wave be *in phase* or *180° out of phase* with the incident wave (*i.e.,* is there a phase change upon transmission)?

2. Reflection at the second boundary

    a. Continue the transmitted ray (from part 1) through the film to the second boundary (on the right). Then draw rays that correspond to the light that is transmitted and reflected at the second boundary.

    b. For light incident on the second boundary, would the reflection at this boundary be more like reflection from a *fixed end* or from a *free end?* Explain.

        At the second boundary, would the reflected wave be *in phase* or *180° out of phase* with the incident wave (in the film)?

3. Transmission at the first boundary

    Continue the reflected ray from part 2 through the film back to the first boundary. Then draw rays that correspond to the light that is transmitted and reflected at this boundary.

    Would there be a phase change on transmission at this boundary?

D. Light of frequency $f = 7.5 \times 10^{14}$ Hz is incident from the left side of the film.

    Determine the numerical values of the:

    • frequency of the wave in film (in Hz)

    • wavelength in air (in nm)

    • wavelength in film (in nm)

E. Suppose that an observer were located on the left side of the film in part C.

    Which of the rays that you drew could reach this observer?

    How would these rays be different if the light were incident at essentially normal incidence?

III. A film of non-uniform thickness

A soap film supported by a vertical loop has settled and is thinner at the top than at the bottom. Light of frequency $f = 7.5 \times 10^{14}$ Hz is incident on the film at essentially normal incidence.

Cross-sectional side view (not to scale)

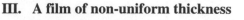

Observer *C*

Observer *A*   Air | Air

Observer *B*

Thinnest part of
soap film ($n = {}^4/_3$)

Thickest part
of soap film

A. Observer *A* is looking at the part of the film that is 75 nm thick.

Consider two reflected rays that reach observer *A,* similar to the rays that you identified in part E of section II.

1. How much farther does one of these rays travel than the other in reaching observer *A?*

2. What is the phase difference between these rays? (Be sure to take into account the phase changes that you identified in part C of section II as well as any phase difference due to path length difference.)

3. Is observer *A* looking at a region of *maximum brightness, minimum brightness,* or *neither?* Explain your reasoning.

B. Observer *B* is looking at the part of the film that is 150 nm thick.

Is this observer looking at a region of *maximum brightness, minimum brightness,* or *neither?* Explain your reasoning.

Tutorials in Introductory Physics
McDermott, Shaffer, & P.E.G., U.Wash.

©Prentice Hall
Preliminary Edition, 1998

C. Observer *C* is looking at the thinnest part of the film, where the film is *extremely* thin. To this observer, would the film appear *bright* or *dark?* Explain your reasoning.

D. Describe the appearance of the film as a whole.

⇨ Check your answers to parts A–D with a tutorial instructor.

E. What are the three smallest film thicknesses for which there would be maximum *constructive* interference?

What are the three smallest film thicknesses for which there would be maximum *destructive* interference?

F. The thickness of the film is 1650 nm at the bottom of the film, where the film is the thickest.

Would this part of the film appear *bright, dark,* or *in between?* Explain.

G. Suppose that the frequency of the incident light were increased.

How, if at all, would the appearance of the thinnest part of the film change?

Would the number of bright and dark regions *increase, decrease,* or *stay the same?* Explain your reasoning.

I. Polarization of light

A. Look at the room lights through one of the polarizing filters provided.

Describe how the filter affects what you see. Does rotating the filter have an effect?

B. Hold a second polarizing filter in front of the first, and look at the room lights again.

Describe how the filter affects the light that you see. How does rotating one of the filters affect what you see in this case?

On the basis of your observations so far, why is it appropriate to use the term *filter* to describe these pieces of apparatus?

How is the behavior of the polarization filters *different* from the behavior of colored acetate filters?

You have learned that light may be thought of as a wave consisting of oscillating electric and magnetic fields. If the electric field in all parts of a light beam oscillates along a single axis, the light beam is said to be *linearly polarized,* or simply, *polarized.* For example, the diagram at right represents a polarized light wave moving in the *x*-direction in which the electric field oscillates only

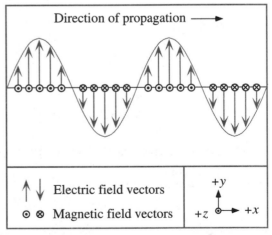

along the *y*-axis. By convention, the direction along which the electric field oscillates (in this case, the *y*-direction) is called the *direction of polarization* of a light beam. If the electric field oscillates in different, random directions within the same light beam, that beam is said to be *unpolarized.*

Tutorials in Introductory Physics
McDermott, Shaffer, & P.E.G., U.Wash.

©Prentice Hall
Preliminary Edition, 1998

II. Polarizing filters

The light transmitted by a polarizing filter (or *polarizer*) depends upon the relative orientation of the polarizer and the electric field in the light wave. Every polarizer has a *direction of polarization,* which is often marked by a line drawn on it. The electric field of the transmitted wave is equal to the component of the electric field of the incident wave that is *parallel* to the direction of polarization of the polarizer.

A. Do the room lights produce polarized light? Explain how you can tell from your observations so far.

B. Suppose that you had two marked polarizers (*i.e.,* their directions of polarization are marked).

Predict how you should orient the polarizers with respect to one another so that the light transmitted through the polarizers would have (1) *maximum* intensity or (2) *minimum* intensity. Discuss your reasoning with your partners and then check your predictions.

When two polarizers are oriented with respect to each other such that the light transmitted through them has minimum intensity, the polarizers are said to be *crossed.*

C. Suppose that you had a polarizer with its direction of polarization marked. How could you use this polarizer to determine the direction of polarization of another (unmarked) polarizer? Explain your reasoning.

D. A beam of light is incident on a polarizer, as shown in the side view diagram below. The light is polarized in a direction that is tilted by an angle θ with respect to the polarizer's direction of polarization. (See front view diagram below.) The vector, $\vec{E}_o$, shown in the diagram represents the amplitude of the electric field of the incident light at one instant at a particular point in space. The corresponding magnetic field vector, $\vec{B}_o$, which is not shown, represents the amplitude of the magnetic field of the incident light.

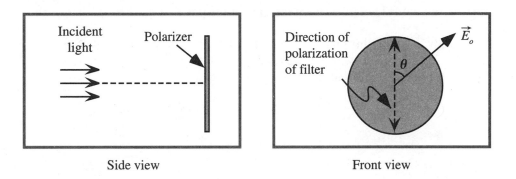

Side view   Front view

Resolve the vector $\vec{E}_o$ into two components (1) the component of the electric field that is *transmitted* by the polarizer and (2) the component of the electric field that is *absorbed* by the polarizer.

What is the *direction* of the electric field that is transmitted by the polarizer? How, if at all, is it different from the direction of the electric field of the incident light? Explain.

Write an expression for the *magnitude* of the electric field of the transmitted light, in terms of E_o and θ.

Write an expression for the *magnitude* of the magnetic field of the transmitted light, in terms of B_o and θ. Explain your reasoning.

Write an expression for the *intensity* of the transmitted light in terms of I_o, the intensity of the incident light, and θ. Show all work. (*Hint:* If the amplitude of the electric field were halved, how would the intensity change?)

⇨ Check your results from part D with a tutorial instructor.

E. An observer is looking at a light source through two polarizers as shown in the side view diagram at right. The polarizers are crossed, that is, they are oriented so that the light transmitted through them has *minimum* intensity.

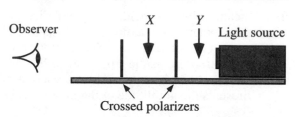

Crossed polarizers

1. Suppose that a third polarizer were inserted at the position marked *X*, shown above.

   Predict how, if at all, this change would affect the intensity of the light reaching the observer. Does your answer depend on the orientation of the third polarizer? Discuss your reasoning with your partners.

   Check your prediction experimentally. (Ask a tutorial instructor to show you the equipment that you need in order to do so.) If your prediction was incorrect, identify those parts of your prediction that were wrong.

   How can you apply your results from part D to help you account for your observations? Support your answer with one or more diagrams.

2. Suppose that instead a third polarizer were inserted at the position marked *Y*, shown above.

   Predict how, if at all, this change would affect the intensity of the light reaching the observer. Does your answer depend on the orientation of the third polarizer? Discuss your reasoning with your partners.

F. Consider a beam of *unpolarized* light that is incident on a polarizer. What is the intensity of the transmitted light in terms of I_o, the intensity of the incident light? (*Hint:* We can think of unpolarized light as equal amounts of light that are polarized *parallel* and *perpendicular* to the direction of polarization of the polarizer.)

Tutorials in Introductory Physics
McDermott, Shaffer, & P.E.G., U.Wash.

©Prentice Hall
Preliminary Edition, 1998

Credits:

Page 93:   Source: *Physics by Inquiry* by Lillian C. McDermott and the Physics Education Group. © 1996 by John Wiley & Sons, Inc. Reprinted by permission.

Page 95:   Source: *Physics by Inquiry* by Lillian C. McDermott and the Physics Education Group. © 1996 by John Wiley & Sons, Inc. Reprinted by permission.

Page 131: Source: *Physics by Inquiry* by Lillian C. McDermott and the Physics Education Group. © 1996 by John Wiley & Sons, Inc. Reprinted by permission.

Page 138: Source: *PSSC Physics* by Uri Haber-Schaim, Judson B. Cross, John H. Dodge, and James A. Walter. © 1971 and 1976 by Education Development Center, Inc. Reprinted by permission.

Page 139: Source: *PSSC Physics* by Uri Haber-Schaim, Judson B. Cross, John H. Dodge, and James A. Walter. © 1971 and 1976 by Education Development Center, Inc. Reprinted by permission.

Page 140: Source: *PSSC Physics* by Uri Haber-Schaim, Judson B. Cross, John H. Dodge, and James A. Walter. © 1971 and 1976 by Education Development Center, Inc. Reprinted by permission.

Page 141: Source: *PSSC Physics* by Uri Haber-Schaim, Judson B. Cross, John H. Dodge, and James A. Walter. © 1971 and 1976 by Education Development Center, Inc. Reprinted by permission.

Page 143: Source: *PSSC Physics* by Uri Haber-Schaim, Judson B. Cross, John H. Dodge, and James A. Walter. © 1971 and 1976 by Education Development Center, Inc. Reprinted by permission.

Page 157: Source: *Physics by Inquiry* by Lillian C. McDermott and the Physics Education Group. © 1996 by John Wiley & Sons, Inc. Reprinted by permission.

Page 161: Source: *Physics by Inquiry* by Lillian C. McDermott and the Physics Education Group. © 1996 by John Wiley & Sons, Inc. Reprinted by permission.

Page 169: Source: *Physics by Inquiry* by Lillian C. McDermott and the Physics Education Group. © 1996 by John Wiley & Sons, Inc. Reprinted by permission.

Page 176: Source: *Physics by Inquiry* by Lillian C. McDermott and the Physics Education Group. © 1996 by John Wiley & Sons, Inc. Reprinted by permission.

Page 186: Source: Film loop *Interference of Waves*. © 1964 by Education Development Center, Inc. Reprinted by permission.

Page 189: Source: *Atlas of Optical Phenomena* by Michel Cagnet, Maurice Françon, and Jean Claude Thrierr. © 1962 by Springer Verlag. Reprinted by permission.

Page 190: Source: *Atlas of Optical Phenomena* by Michel Cagnet, Maurice Françon, and Jean Claude Thrierr. © 1962 by Springer Verlag. Reprinted by permission.

Page 191: Source: *Atlas of Optical Phenomena* by Michel Cagnet, Maurice Françon, and Jean Claude Thrierr. © 1962 by Springer Verlag. Reprinted by permission.

Page 194: Source: *Atlas of Optical Phenomena* by Michel Cagnet, Maurice Françon, and Jean Claude Thrierr. © 1962 by Springer Verlag. Reprinted by permission.

Page 197: Source: *PSSC Physics* by Uri Haber-Schaim, Judson B. Cross, John H. Dodge, and James A. Walter. © 1971 and 1976 by Education Development Center, Inc. Reprinted by permission.

Page 198: Source: Vincent Mallette, Georgia Tech., Atlanta, Georgia.

Page 200: Source: K. Hendry, Seattle, Washington.

Page 201: Source: *Atlas of Optical Phenomena* by Michel Cagnet, Maurice Françon, and Jean Claude Thrierr. © 1962 by Springer Verlag. Reprinted by permission.